THE CHEMISTRY OF FIREWORKS

RSC Paperbacks

RSC Paperbacks are a series of inexpensive texts suitable for teachers and students and give a clear, readable introduction to selected topics in chemistry. They should also appeal to the general chemist. For further information on available titles contact:

Sales and Customer Care Department, Royal Society of Chemistry,
Thomas Graham House, Science Park, Milton Road, Cambridge CB4 0WF, UK
Telephone: +44 (0) 1223 432360; Fax: +44 (0) 1223 423429; E-mail: sales@rsc.org

New Titles Available

Food – The Chemistry of its Components (Third Edition)
by T. P. Coultate
Archaeological Chemistry
by A. M. Pollard and C. Heron
The Chemistry of Paper
by J. C. Roberts
Introduction to Glass Science and Technology
by James E. Shelby
Food Flavours: Biology and Chemistry
by Carolyn Fisher and Thomas R. Scott
Adhesion Science
by J. Comyn
The Chemistry of Polymers (Second Edition)
by John W. Nicholson
A Working Method Approach for Introductory Physical Chemistry Calculations
by Brian Murphy, Clair Murphy and Brian J. Hathaway
The Chemistry of Explosives
by Jacqueline Akhavan
Basic Principles of Inorganic Chemistry – Making the Connections
by Brian Murphy, Clair Murphy and Brian J. Hathaway
The Chemistry of Fragrances
compiled by David Pybus and Charles Sell
Polymers and the Environment
by Gerald Scott
Brewing
by Ian S. Hornsey
The Chemistry of Fireworks
by Michael S. Russell

Future titles may be obtained immediately on publication by placing a standing order for RSC Paperbacks. Information on this is available from the address above.

RSC Paperbacks

THE CHEMISTRY OF FIREWORKS

MICHAEL S. RUSSELL

ROYAL SOCIETY OF CHEMISTRY

ISBN 0-85404-598-8

A catalogue record for this book is available from the British Library

Published by The Royal Society of Chemistry,
Thomas Graham House, Science Park, Milton Road,
Cambridge CB4 0WF, UK

For further information see our web site at www.rsc.org

Typeset by Paston Prepress Ltd, Beccles, Suffolk, NR34 9QG
Printed by Athenaeum Press Ltd, Gateshead, Tyne & Wear, UK

Preface

This book does not claim to be a definitive text on fireworks and the fireworks industry. It is primarily an introduction to the basic science of fireworks with particular emphasis on the underlying chemistry and physics.

The historical material – stemming from several well known sources – is valuable for its technical content. The subject matter then advances to a presentation on the characteristics of gunpowder, whose unique properties cause it to be the mainstay of the fireworks industry, even today.

Succeeding chapters describe the manufacture and functioning of the most popular fireworks, including rockets, shells, fountains, roman candles, bangers, gerbs and wheels in what is hoped is a stimulating and easily assimilated way for those approaching the subject for the first time.

Whilst the book is aimed at students with A-level qualifications, or equivalent, it is also intended to be useful background material and a source of reference for anybody engaged in a study of pyrotechnics as applied to fireworks. Chapters on fireworks safety and legislation complete the book and deserve special mention.

None of this book could have appeared – at least as a commercial project – without the enthusiastic advice and very willing co-operation of Mr John Stone and the late Gordon Curtis of Pains-Wessex Ltd. I am also indebted to Bill Deeker of Pains Fireworks for permission to reproduce the black and white photographs, and to David Cox who gave advice on their selection and indeed took many of the pictures.

Probably the most comprehensive general text on fireworks is the book by the Rev. Ronald Lancaster and co-contributors (*Fireworks – Principles and Practice*, 3rd Edition) and I am grateful to him for discussing my book with me and for giving permission to quote some of his formulae.

I must also acknowledge the many friends and display operators with whom I have enjoyed sharing my passion for fireworks over the years. These include Ray Harrison, Henry Dunlop, Campbell Wilson, Chris Wilson, Ken Norton, Dave Laurence, Andy Goodwin and Debbie, Jonathon Webb, Steve Cornall, Dale Sullivan, Roly Harrison, the late

Wally Betts, Kevin Russell, Robert Stevens and my number two daughter, Jane.

My grateful acknowledgements also go to Kay, Carol and Peter who worked wonders on the computer.

Finally, I would like to thank my dear wife, Lyn, who sat through countless evenings of researching, writing and re-writing.

But having a degree in chemistry and a birthday on the 5th November, what else could a person choose to write about?

Michael S. Russell

Contents

Chapter 12
Fireworks Legislation   103

Glossary

AFTERGLOW The glowing remains produced by the firing of gunpowder-based products such as quickmatch. It is very important that any afterglow is extinguished, especially when reloading shells or mines into mortar tubes.

AMORCE A toy cap that consists of a paper envelope containing explosive composition and which forms part of a roll.

APOGEE The point at which a projectile, such as a rocket, is at its greatest height above the Earth.

BANGER A small tube containing gunpowder that is ignited from a simple fuse.

BATTERY A group of Roman candles or a set of similar or connected fireworks.

BINDER A substance such as varnish, shellac or gum arabic that is used to bind together the components of a pressed composition.

BLACK POWDER (synonymous with Gunpowder) An intimately milled mixture of potassium nitrate, sulfur and charcoal that has propellant or explosive properties.

BOMBETTE A combination of candles and/or shells packed in a box and fired by interconnecting fuse.

BOUQUET Simultaneously ejected coloured stars from rockets or shells.

BRITISH STANDARD (Fireworks) BS7114: Part 1 Classification of Fireworks, Part 2 Specification for Fireworks, Part 3 Methods of Test for Fireworks (BSI Sales Dept, Linford Wood, Milton Keynes MK14 6SL, UK).

BURNING RATE   The regression of a reaction zone of a pressed composition or fuse, usually expressed in millimetres per second. The volume burning rate is expressed in cubic centimetres per second while the mass burning rate, which is the product of the composition density and the volume burning rate, is expressed in grams per second.

BURSTER   Explosive composition which will burn to evolve gas which in turn is intended to burst open the firework case in order to expel one or more pyrotechnic units.

CAP   Small amount of impact-sensitive explosive composition contained in a non-metallic envelope.

CATHERINE WHEEL   A firework consisting of spiral of pyrotechnic composition in a paper case that rotates on a pin.

CHINESE FIRE   A pyrotechnic composition based on meal powder, iron filings and charcoal which is designed to produce visual effects including sparks.

CLER   The Classification and Labelling of Explosives Regulations 1983. HMSO, HS(R)17, ISBN 0 11 88 3706 0.

CLOVE HITCH   A knot which is used (or should be used) by fireworks operators in order to effectively secure stakes to posts, *etc*. It is affected by looping the cord around the timber so that the loop crosses over at the front. The top section of cord is then looped again around the timber, but this time passed **underneath** the cross to form a double loop. The knot is then pulled taut and secured with a half-granny knot or similar.

COMET   A single large star expelled from a firework such as a mine.

COMPOSITION (Explosive)   An intimate mixture of fuels, oxidisers and additives of such particle size that, when pressed, it is capable of producing pyrotechnic effects.

CONSUMER PROTECTION   The Fireworks (Safety) Regulations 1997. SI No. 2294.

CRACKER SNAP   Two overlapping strips of paper or card with a friction-sensitive explosive composition in contact with an abrasive surface.

CROWN WHEEL   A circular firework with a central pivot that rests upon a nail or spike to give the effect of spinning like a wheel and then rising into the sky.

DEBRIS   Any part of a firework that remains after the firework has finished functioning.

DEFLAGRATION   A burning surface reaction that progresses as successive layers are raised to their initiation temperature and ignited.

DELAY TRAIN   A combination of igniters and fuses that burn for a predetermined time before igniting the main explosive composition.

DEVICE   An assembly consisting of various type of fireworks, linked together, each producing specific pyrotechnic effects, with a single point of ignition.

EQUATION (chemical)   A representation of a chemical reaction, using the symbols of the elements to represent the actual atoms and molecules taking part in a reaction. For example, a classical, but simplified, overall reaction for the deflagration of gunpowder is as follows:

$$2KNO_3 + S + 3C \rightarrow K_2S + 3CO_2 + N_2$$

Potassium nitrate   Sulfur   Charcoal   Potassium sulfide   Carbon dioxide   Nitrogen

EXOTHERMIC   A term used to describe a chemical reaction in which energy in the form of heat is released.

EXPLOSION   A violent and rapid increase of pressure in a confined space, brought about, in fireworks, by internal exothermic chemical reactions in which relatively large volumes of gases are produced.

EXPLOSIVE   Substances which undergo rapid chemical changes, with the (**chemical**) production of gases, on being heated or otherwise initiated.

EXPLOSIVES ACTS   Acts of Parliament whose objectives are to control the manufacture, keeping, sale, conveyance, importation and criminal use of explosives.

FIRECRACKER   An early form of banger using gunpowder and paper cases commonly tied into bundles or strips.

FIREWORK   An article containing an explosive composition which, upon functioning, will burn and/or explode to produce visual and/or aural effects which are used for entertainment or signalling.

FLASH POWDER   A mixture of fuels, oxidisers and other additives that is capable of being initiated to undergo fast deflagration which is usually accompanied by smoke and a bright flash.

FLITTER (colloquial)   A spark that gives a transient but twinkling effect.

FOUNTAIN   A long, tubular firework from which a jet or spray of sparks issues, sometimes accompanied by stars.

FUEL   Any substance capable of reacting with oxygen and oxygen carriers (oxidisers) with the evolution of heat.

FUSE   An item with pyrotechnic or explosive components that is intended to be ignited in order to start the firework functioning or to transmit ignition from one part of a firework to another.

FUSEE   An article resembling a safety match but which has additional pyrotechnic composition that glows after ignition and is essentially wind-proof and weather-proof. Used for lighting fuses.

GERB   A small tubular firework from which a jet or spray issues.

GRAIN   The particulate matter of a granulated composition, or a charge of solid rocket propellant, or a unit of mass, where one grain = 0.0648 g.

GUNPOWDER   See BLACK POWDER.

HSE   Health and Safety Executive. (Explosives Inspectorate. Magdalen House, Bootle, Merseyside L20 3QZ, UK).

IGNITION   Initiating combustion by raising the temperature of the reactants to the ignition temperature.

INCANDESCENCE   The emission of light caused by high temperatures; white or bright red heat.

INITIATOR   The first component in a pyrotechnic or explosive train.

LANCE   Small, tubular firework designed to emit a coloured flame for about 90 seconds. Used for a visual effect in set-pieces.

LATTICE   A framework of crossed laths used for supporting lances or other fireworks.

LOCAE   List of Classified and Authorised Explosives. (HSE Books, PO Box 1999, Sudbury, Suffolk, CO10 6FS, UK).

MAROON   A firework that is fired from a mortar and explodes with a loud report. Usually used to signal the start of a display.

MEAL POWDER   A very fine particle-size gunpowder that is used for priming and making matches.

MILL   The apparatus for reducing the particle size of pyrotechnic ingredients and/or intimately mixing the said ingredients.

MINE   A firework that is fired from a mortar and which contains a single propellant charge and pyrotechnic units.

MIS-FIRE   The failure of any firework or pyrotechnic unit to function within the expected time. In the event of a mis-fire, at least 20 minutes should be allowed before approaching the firework.

MORTAR   A tube from which a mine or shell may be fired.

NEC   Net Explosive Content. The mass of explosive composition within a firework.

OPERATOR   A person who operates fireworks or pyrotechnic displays (usually in conditions of darkness, cold and damp, with little financial reward, but inestimable dedication).

PARTY POPPER   Small hand-held firework operated by a pull-string.

PAYLOAD   The total mass of pyrotechnic effects carried by a rocket, *etc.*

PEC   The Packing of Explosives for Carriage Regulations 1991 (SI 1991/2097) HMSO, ISBN 0 11 015 197 X.

PIC   Plastic Igniter Cord. A fuse burning with an intense flame progressively along its length. Used for igniting the match attached to shells, *etc.* Different burning speeds are available from $49\,\mathrm{s\,m^{-1}}$ to $3.3\,\mathrm{s\,m^{-1}}$.

PIPED MATCH   A fuse consisting of quickmatch enclosed in a paper pipe that serves to increase the burning speed.

PORTFIRE   A hand-held tubular appliance containing slow-burning explosive composition which will emit a small flame. Commonly used for lighting fuses.

PRIMING   A layer of readily ignited explosive composition that is applied to the surface of the main composition in order to facilitate ignition.

PROPELLANT   An explosive composition that burns with the characteristics necessary for propelling shells and rockets, *etc.*

PYROTECHNIC   Of, like, or relating to fireworks.

PYROTECHNIST   A person skilled in the art of making or using fireworks.

QUICKMATCH   A fuse consisting of gunpowder coated onto cotton yarn using an adhesive such as gum arabic.

RAD HAZ   Radio Hazard. The hazard associated with the use of electro-explosive devices (EEDs) such as wirebridge fuseheads in the vicinity of radio-frequency transmitting equipment (BS 6657: Prevention of Inadvertent Initiation of Electro-explosive Devices by Radio-frequency Radiation).

ROCKET   A self-propelled firework with stick for stabilisation of flight.

ROCKET LAUNCHER   A tube, frame, box or base from which rockets may be launched.

ROMAN CANDLE   A tubular firework usually containing a plurality of alternate pyrotechnic units and propellant charges.

SAXON   A tubular firework containing a pair of opposing nozzles that is designed to rotate by virtue of a central pivot.

SERIES FIRING   A method of firing fireworks electrically by connecting wirebridge fusehead igniters in series, *i.e.* one after the other, so that the current flows through each in turn.

SERPENT   The preformed shape of explosive composition or an integral container of explosive composition that functions with the emission of expanded residue.

SET-PIECE   An assembly consisting of lances and other fireworks linked together to form images and other pyrotechnic effects with a single point of ignition.

SHELL   A firework designed to be projected from a mortar tube and containing propellant charge, delay fuse, burster and pyrotechnic units.

SPARKLER   Wire coated along one end with explosive composition, and designed either to be non-hand held (*i.e.* free-standing or fixed to a support) or to be held in the hand.

STAR   A compressed pellet of explosive composition designed to be projected as a pyrotechnic unit, with visual effects.

SQUIB   A small tubular firework, containing gunpowder, that makes a hissing sound and then explodes.

STOICHIOMETRIC MIXTURE   A balanced mixture which, on reaction, will yield a stoichiometric compound. For example, two molecules of hydrogen and one molecule of oxygen constitute a stoichiometric mixture because they yield exactly two molecules of water on combustion. Such a balance is important when formulating pyrotechnic compositions.

THROWDOWN   An article containing an impact-sensitive explosive composition.

TOURBILLIONS   Fireworks designed to revolve on the ground and then ascend. They consist of tubes with opposing nozzles and small wings. They used to be known as 'aeroplanes' in the UK.

WHEEL   Any firework that is designed to rotate about a fixed point.

WHISTLER   A firework designed to whistle by virtue of an explosive composition containing gallic acid or similar.

WIREBRIDGE FUSEHEAD   An electric igniter containing a bridgewire surrounded by a small bead of explosive composition designed to emit a short burst of gas and flame.

Chapter 1

Historical Introduction

EARLY INCENDIARY DEVICES

Working with fire probably began about half a million years ago when patriarchal cavemen realised that they felt the cold and began rubbing pieces of wood together until the friction caused an ignition. In fact, it is none too easy to generate fire in this way but we have all seen contrivances driven by coils of leather that spin a pointed stick against a wooden notch until it smokes and eventually bursts into flame.

Now it was originally thought that fire was a kind of substance and that this substance generated flames when it met the air. It is only within the last 200 years or so that fire was correctly interpreted as being a form of energy where the flames are defined as regions of luminous hot gas.

To find evidence of the first application of fire in the creation of 'special effects' it is necessary to go back some 1400 years when the naturally-occurring substances petroleum and naphtha were employed by the Greeks as an early form of napalm. In the characteristically unfriendly practices of those times, one Kallinikos from Heliopolis of Syria set forth in armed conflict against the Arabs. He had equipped fast-sailing galleys with cauldrons of what amounted to burning crude oil and proceeded to set the boats of the enemy ablaze, with the men still aboard. The incendiary was called 'Greek Fire'.

The ploy must have worked because the subsequent narrative tells us that the Byzantines then capitalised on their secret weapon by the wholesale destruction of the Moslem fleet at Cyzicus and continued to win naval battles in this way for several centuries afterwards.

DEVELOPMENT OF BLACK POWDER

By about the eighth century AD, Chinese alchemists, amongst others, were preoccupied with discovering the elixir of life. Concoctions were made containing all manner of substances including oils, honey and

1

beeswax, but among the most significant, so far as future firework makers were concerned, were the ingredients sulfur and saltpetre. Unbeknown to the ancients, their brew of honey, sulfur and saltpetre (potassium nitrate) was special in that, on evaporation over heat, the contents would suddenly erupt into a wall of flame. By chance, the experimenters had produced the exact proportions by which the molten sulfur and what was left of the honey were acting as fuels that were subsequently oxidised by the oxygen from the potassium nitrate in what is now known as an 'exothermic chemical reaction', and a fairly vigorous one at that! In purified form, the chemicals sulfur and saltpetre are used to this day in what is without doubt the most important tool of the firework makers, *i.e.* gunpowder.

These dangerous early experiments led to many secret or banned recipes, but enough information was disseminated to enable the details of the discovery to be brought to Europe. However, the place and date of the invention of true gunpowder are still unknown and have been the subject of extensive but inconclusive investigation.

Once the reactive tendencies of potassium nitrate were unleashed it was simply a matter of time before the third vital ingredient, charcoal, was added to complete the famous gunpowder recipe of charcoal, sulfur and potassium nitrate. Needless to say, much time and effort were expended before the alchemists produced a successful product.

As with many notable inventions, the credit for the discovery is usually coloured by patriotism, each country putting forward its own 'inventor'. What is significant, however, is that by about 1000AD the Chinese were using a propellant similar to gunpowder in crude forms of rockets (Flying Fire), together with grenades and even toxic smokes. For example, a recipe in the Wu Ching Tsung Yao dated 1044 describes a mixture containing sulfur, saltpetre, arsenic salts, lead salts, oils and waxes to give a toxic incendiary that could be launched from a catapult.

More peaceful uses of these crude articles appeared in the form of 'fire crackers' – the first fireworks? One mixture corresponded quite closely to modern gunpowder in that it contained saltpetre, sulfur and willow charcoal. The 'fire cracker' was said to consist of a loosely-filled parchment tube tied tightly at both ends and with the introduction of a small hole to accept a match or fuse. All of these incendiary mixtures, presumably containing saltpetre, are mentioned in Chinese work dating from the eleventh century AD. Thus, in theory at least, the Battle of Hastings could have been one of 'Greek Fire', incendiary rockets and grenades.

Skipping about two centuries, the activities of one experimenter typify the development of early black powder. His work took place between about 1235 and 1290AD and he is reputed to have been the first scholar in Northern Europe who was skilled in the use of black powder. In essence, his work provided the backbone of all early chemical purification and formulation, without which the development of true gunpowders would not have been possible. His name was Roger Bacon (Figure 1.1).

Born in about 1214, Bacon became a monk but was educated at Oxford before gaining a doctorate in Paris. His subjects included philosophy, divinity, mathematics, physics, chemistry and even cosmology. He carefully purified potassium nitrate (by recrystallisation from water) and went on to experiment with different proportions of the other two ingredients (sulfur and willow charcoal) until he was satisfied that,

> *By the flash and combustion of fires, and by the horror of sounds, wonders can be wrought, and at any distance that we wish, so that a man can hardly protect himself or endure it.*

Of course, 'The Church' was not wildly enthusiastic with the prospect of one of its disciples practising such fiendish alchemy, and Bacon served ten years' imprisonment. But he preserved his most famous recipe of *ca.* 1252AD in the form of an anagram, which on deciphering reads 'of saltpetre take six parts, five of young willow (charcoal) and five of sulfur and so you will make thunder and lightning'. In percentage terms, the 6:5:5 formula translates as saltpetre 37.50 parts by weight, charcoal 31.25 and sulfur 31.25 parts.

In fact, Roger Bacon's formula was not too dissimilar from early Chinese recipes. But being natural products, all three ingredients were

Figure 1.1 *Roger Bacon*

of variable purity. For example, the crude Indian or Chinese saltpetre was richer in true saltpetre than the European material, but all required recrystallisation. The preferred process seems to have involved wood ashes, containing potassium carbonate, which precipitated deliquescent calcium salts from the saltpetre solution. The solution was then passed through a filter, boiled to reduce the volume of water and then left until the transparent plates of purified saltpetre were formed.

Sulfur occurs widely in nature as the element and was thus easily obtainable by the ancients. The Chinese had rich natural deposits, and the substance is readily purified by sublimation, a process in which the native sulfur is heated and the evolved vapour collected directly as a pure solid.

Charcoal was made from common deciduous woods such as birch, willow or alder, the last two being preferred.

The wood is simply carbonised at relatively low temperatures in a restricted air supply to form an amorphous, quasi-graphitic carbon of very fine particle size. Although of reasonably high purity, it is the enormous surface area per unit mass of the charcoal which makes it very adsorbent to water vapour, and this property is conferred to the black powder mix, as Roger Bacon would have soon realised.

Guns were invented shortly after Bacon's death in about 1292 and so he never used the term 'gunpowder'. However, he had certainly had experience of fireworks for which his early black powder recipe would have been perfectly suitable. In the *Opus Majus* he wrote:

> *We have an example of this in that toy of children which is made in many parts of the world, namely an instrument as large as the human thumb. From the force of the salt called saltpetre so horrible a sound is produced at the bursting of so small a thing, namely a small piece of parchment that we perceive it exceeds the roar of sharp thunder, and the flash exceeds the greatest brilliancy of the lightning accompanying the thunder.*

In experimenting with fireworks, Roger Bacon and other medieval chemists discovered that a loose, open tray of powder was all that was needed to produce a flash, but in order to produce the bang the powder needed to be confined, and this has great significance. And even with his unbalanced 6:5:5 formula, Bacon was able to deduce these fundamental ballistic effects.

This short introduction to gunpowder would not be complete without reference to its final development and one or two subsequent events that were to change the course of history.

In lighting a firework we are going back at least 1000 years. The potassium nitrate in the blue touch-paper or the match burns in much the same way as it did when the Arabs or the Chinese played with their fire crackers. The smell of the sulfur when it forms hydrogen sulfide on combustion would have been much the same, as would the dense white smoke that is so characteristic of gunpowder. But modern fireworks are reliable products. The gunpowder has a consistent burning rate and is less affected by moisture than it would have been in the eleventh century. Obviously it was in the interests of the future markets that the experimenters persevered, and their pioneering work was by no means trivial.

First, true gunpowder is not just a 'loose' mixture of potassium nitrate, sulfur and charcoal. Indeed, if the three ingredients are mixed in this way then a greyish powder results that is almost impossible to light. If ignition does occur the burning is fitful and prone to extinguishment. In order to overcome these deficiencies the ingredients must be brought into intimate contact. The charcoal and sulfur are milled together with 2–3% of water in a tumbling barrel, then the potassium nitrate is added and the damp mixture is further milled under rollers before being pressed into a cake using a hydraulic press at a pressure of about 2 tonnes.

As with the modern fireworks industry, pressing is preferred over more forceful techniques, but even so, fires regularly break out in presses. Milling is not without hazard either, especially when the large wheel mills weigh several tonnes and the powder batch is around 150 kg.

After pressing, the gunpowder cake is broken and this corning or granulating is the most dangerous of all manufacturing operations.

It is necessary to crack the cake between crusher rolls to form the grains (see Figure 1.2) which are subsequently graded by sieving. The 'finishing' process involves tumbling and drying the granulated powder

Figure 1.2 *Gunpowder*

in wooden barrels in the presence of graphite to give a polished or glazed appearance. The granulated and glazed gunpowders were found to be more moisture-resistant than the early fine powders and the ignition and burning consistency was also much improved. It is the 'fines' or corning mill dust that is used in fuse powder and by the makers of fireworks.

Of course, in the Middle Ages the emerging gunpowder industry relied on mortars and pestles to do the mixing, and the recipes were changed in what was, in reality, an enrichment of the saltpetre content to give faster burning and ever more powerful powders for yet another historically important invention – the gun.

Thus, in Arabic work dating between perhaps 1300 and 1350AD, gunpowder is described as a propellant. Cannon were also known in Europe by that time and were used in the defence of castles and villages. In 1338, cannon and powder were provided for the protection of the ports of Harfleur and L'Heure against Edward III. From about 1340 onwards there is frequent mention of the use of guns, and by 1400 the Crown in England possessed a stock of guns and gunpowder.

It is interesting to record how the composition of gunpowder changed as history progressed (Table 1.1) and how the 75:15:10 mix of 1781 remains in use to the present day.

In fact, most of the improvements to gunpowders after about 1600AD concerned the methods of manufacture, there being no question that the proportions of the three components were correctly balanced for chemical reaction, that is to say 'stoichiometric'.

An approximate equation for the burning of black powder has been given as in reaction (1.1).

$$74KNO_3 + 96C + 30S + 16H_2O \longrightarrow 35N_2 + 56CO_2 + 14CO + 3CH_4$$
$$+ 2H_2S + 4H_2 + 19K_2CO_3 + 7K_2SO_4$$
$$+ 8K_2S_2O_3 + 2K_2S + 2KSCN$$
$$+ (NH_4)_2CO_3 + C + S \qquad (1.1)$$

The above reaction corresponds to a composition containing saltpetre (75.7%), charcoal (11.7%), sulfur (9.7%) and moisture (2.9%).

Table 1.1 *Examples of gunpowder compositions*[a]

Date	*ca.* 1252	*ca.* 1350	*ca.* 1560	*ca.* 1635	*ca.* 1781
Saltpetre	37.50	66.6	50.0	75.0	75
Charcoal	31.25	22.2	33.3	12.5	15
Sulfur	31.25	11.1	16.6	12.5	10

[a] Compositions given in parts by weight.

APPLICATION OF BLACK POWDER TO FIREWORKS

The fireworks industry also benefited from these improvements, which was reflected in the growing popularity of organised displays and the diversity of the pyrotechnic effects so presented.

Historically, it is generally accepted that the first fireworks were developed in far-eastern countries, notably India or China, for display at religious festivals, and that knowledge of the art subsequently spread to Europe, probably *via* the Arab kingdom. The Italians are credited with introducing the firework industry in Europe, again promoting their use for public occasions before the manufacture was adopted by neighbouring countries such as France and Germany. By the sixteenth century, fireworks displays were being given in England, and it is documented that Elizabeth I witnessed such an event in August 1572.

Although the early displays in England were enthusiastically received, it must be admitted that most of the pyrotechnic art, and indeed the operators and equipment, originated from Europe – foreign workers were still giving displays in England as late as 1775. It may also be noted, in passing, that in the early seventeenth century the making, purchasing or keeping of fireworks was ruled to be illegal; this was due, in no small measure to the famous (or infamous) attempt to blow up the Houses of Parliament in 1605 by a certain Mr Guy Fawkes using 36 barrels of gunpowder.

The conspiracy is alleged to have begun in 1604 during the second year of the reign of James I, when a group of Catholic fanatics decided that the Establishment must go. Five conspirators, including Guy Fawkes, commenced digging under the main parliamentary building in an attempt to undermine it, and in doing so came across a cellar which was being used by a coal dealer. This they duly filled with 'powder, faggots and billets'. Timing the event to coincide with the State Opening of Parliament on the 5th of November 1605 meant that the conspirators could also claim the life of the King. However, a warning letter was sent to some members of parliament beforehand, and this was read not only by the Secretary of State but also by James I who, with amazing insight, correctly interpreted it as meaning an explosion on November 5th.

The vaults under the main chamber were visited by the Lord Chamberlain on the 4th November and there they found 'a tall and desperate looking fellow' who identified himself as Guido Fawkes. On the 5th of November, magistrates examined the neighbouring house and cellar where they arrested Fawkes who was 'just leaving'.

Guy Fawkes was tortured and his accomplices arrested, tried and executed. The Establishment was clearly not ecstatic about the fact that the plot had so nearly succeeded, and Fawkes was tried at Westminster on 27 January and ceremoniously executed on 30 January 1606.

All of this was subsequently of great benefit to the British fireworks industry, of course, which has capitalised on the 5th of November celebrations ever since. However, any other country in the world might have bent the truth a little and claimed in the history books that the plot took place on a nice, warm day in August rather than in cold and damp November – even if only for the sakes of the fireworks operators!

By the nineteenth century, English firework makers including Brock, Pain and Wells had established themselves in the London area to be later followed by Standard Fireworks and others in the North. Thus the availability of locally-produced gunpowder and fireworks was enough to eschew any drift towards European suppliers.

Although the Explosives Act of 1875 expressly forbids the manufacture of gunpowder or of fireworks outside a licensed factory there have been tales of the Yorkshire miners who, amongst others, produced squibs (small exploding fireworks) by packing gunpowder into rolled paper cases. These were used for blasting but were also said to be effective in clearing the soot from the flues of domestic homes. Under these circumstances it is not difficult to imagine a cottage industry springing up whereby the squibs were turned into crackers or any other simple form of fireworks which were then sold locally.

And so on to the twentieth century when the emergence of free trading between nations once again meant that fireworks, and more importantly, gunpowder, were available from around the world. Gunpowder is no longer manufactured in the United Kingdom and supplies are procured from Spain, Germany, South America and the Far East, as are fireworks. Of the original esteemed group of factories, few survive today and even fewer make any fireworks, relying instead on the magazine storage of imported products to effect their displays, just as they did 300 years ago.

FURTHER USES OF BLACK POWDER

Of gunpowder itself, although it has a long and colourful history, its use as an explosive dwindled into insignificance after the domination enjoyed by the much more powerful high explosives that succeeded it. But besides fireworks, there are still a few 'niche' applications where the unusual burning properties of gunpowder and kindred substances may be exploited. For example, as a 'low explosive' it is suitable for controlled blasting in which the treatment of the stone must be mild. It is therefore used in the manufacture of roofing slates and in quarrying for paving stones, the powder grains being freely poured into boreholes.

It is also employed by the military in making priming charges for smokeless powders. In the largest calibres the gunpowder grains are sewn into quilted silk bags that fit over the ends of the cordite charges to promote ignition. It also finds use in the production of fuses, pyrotechnic

Table 1.2 *Examples of application of black powder*

Application	Composition		
	KNO_3	Charcoal	Sulfur
1 Lift charge or burst charge	75	15	10
2 Priming powder	70	30	0
3 Blasting powder	68	18	14
4 Rocket propellant	62	28	10
5 Delay fire	62	18	20
6 Sparking composition	60	12	28
7 White smoke	50	0	50
8 Fire extinguishing smoke	85	15	0

'special effects', bird-scaring cartridges, cartridge actuators and small-arms 'blanks'.

Apart from its unique property of burning quickly at relatively low confinement it is not prone to detonation. Under normal conditions the maximum rate of explosion is about $500 \, m \, s^{-1}$. In the absence of moisture, gunpowder in also extremely stable. It has been documented that until World War I it was the practice of the French Army to preserve any batches of gunpowder that had proved especially good. These were used in time train fuses and it was claimed that some batches so preserved dated from Napoleonic times.

Perhaps the most unusual modern application of potassium nitrate-based powders is in the fire protection industry. The white smoke mainly consists of potassium carbonate and this has been found to have fire extinguishing properties due to the way in which the potassium salt in the smoke interferes with the combustion chemistry of a fire.

Large grenades containing up to 2 kg of powder composition have been used by European firefighters where, for example, in burning buildings the smoking grenades are simply hurled through the plate glass windows. Now that really is 'fighting fire with fire'!

Table 1.2 lists the current applications of black powder. In general, as the balance of the ingredients in the composition shifts from the near stoichiometric 75:15:10 mix, the rate of burning decreases but is still fast enough to be of major importance in firework rockets, delay fuses, igniters and pyrotechnic smokes.

Chapter 2

The Characteristics of Black Powder

For about 500 years black powder enjoyed dominance as a propellant, explosive and igniter, and as the major firework ingredient its use remains unsurpassed. The prime reason for the longevity of black powder is its 'quickness', even at relatively low pressures, and this is brought about by manufacturing techniques (as described earlier) and the chemical reactivity of the constituents.

For example, the KNO_3 crystals undergo a sharp solid–solid (rhombic to trigonal) transition at 130 °C which has the effect of 'loosening' the solid structure, making the substance more reactive and easier to ignite. This well known transition is referred to as the 'nitrate spin' where the anion has been likened to a three-legged stool, spinning on its vertical axis. The oxygen ions in the nitrate cluster are indistinguishable from each other and this makes rotation easier.

Besides acting as a fuel, the finely-divided charcoal has a very high surface area which means that the substance can absorb appreciable quantities of sulfur dioxide (SO_2) produced during combustion whilst allowing the remaining sulfur to perform more effectively as a fuel.

Sulfur exhibits pseudo-plastic behaviour, especially under pressure, with the result that the compaction of gunpowder during manufacture produces considerable plastic flow forming a conglomerate of interconnecting passageways to the extent that the internal free volume can fall to a few percent. It is this reduced porosity that gives black powder its unique burning properties and makes the material perform almost as a compound rather than as a composite mixture.

INFLUENCE OF PELLET DENSITY ON BURNING TIME

The plastic flow of sulfur is thought to be responsible for the increase in the burning time that is experienced with black powders of various pellet sizes as the pellets (or grains) are compressed to higher bulk densities.

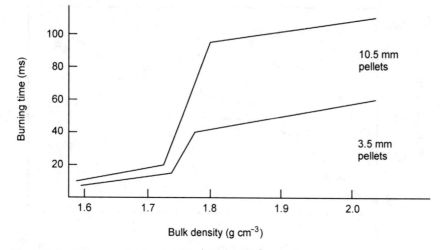

Figure 2.1 *Influence of pellet density on black powder burning time*

Figure 2.1 shows the result of compressing 3.5 mm and 10.5 mm pellets to densities ranging from 1.6 to 2.0 g cm^{-3}. Although the burning process is still not fully understood, the increase of burning time with density is likely to be associated with mechanisms involving the free volume or 'openness' of the pellets.

INFLUENCE OF MOISTURE ON BURNING TIME

Increasing amounts of moisture are also known to produce significant increases in the burning time of gunpowder grains. Water will degrade the performance of most pyrotechnics by virtue of unwanted side reactions, and in the case of gunpowder the adverse effect of moisture is also thought to be as a result of occupying the free volume.

Figure 2.2 shows that an increase in moisture level from 1 to 3% is sufficient to reduce the burning rate by approximately half.

THERMAL DECOMPOSITION

At temperatures below the ignition point, the thermal decomposition of black powder provides an interesting insight into the processes which are thought to control the reaction rate during subsequent burning. In decomposition experiments it has been shown that the overall reaction proceeds in several steps. As the temperature is increased the steps become shorter and the reaction faster. Since these reactions involve gases, the effect of pressure is also important.

The first reaction has been shown to be the formation of hydrogen

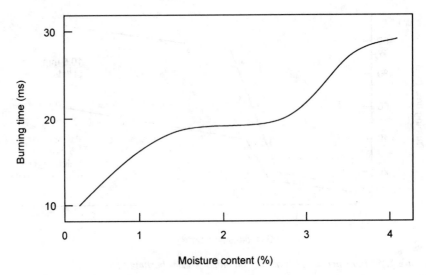

Figure 2.2 *Influence of moisture content on black powder burning time*

sulfide (H_2S) from sulfur and volatile organic material originating from the charcoal as shown in reaction (2.1),

$$S + \text{organic material} \longrightarrow H_2S \tag{2.1}$$

while at the same time,

$$KNO_3 + \text{organic material} \longrightarrow NO_2 \tag{2.2}$$

The NO_2 can also be produced by reactions between sulfur and potassium nitrate (KNO_3) with the formation of nitric oxide (NO) and nitrogen dioxide (NO_2) as shown in reactions (2.3) and (2.4):

$$2KNO_3 + S \longrightarrow K_2SO_4 + 2NO \tag{2.3}$$

$$KNO_3 + 2NO \longrightarrow KNO_2 + NO + NO_2 \tag{2.4}$$

There is then a gas-phase reaction between the main products of these reactions, as in reaction (2.5):

$$H_2S + NO_2 \longrightarrow H_2O + S + NO \tag{2.5}$$

It has been suggested that reaction (2.5), with the regeneration of sulfur, proceeds until all of the H_2S has been used up. The NO_2 then reacts with the free sulfur as in reaction (2.6):

$$2NO_2 + 2S \longrightarrow 2SO_2 + N_2 \tag{2.6}$$

The SO_2 produced in this way then immediately reacts with the KNO_3 as shown in reaction (2.7):

$$2KNO_3 + SO_2 \longrightarrow K_2SO_4 + 2NO_2 \qquad (2.7)$$

The NO_2 is not liberated at this stage but continues the chain. Reactions (2.5) and (2.6) are endothermic (absorb heat) but reaction (2.7) is strongly exothermic and produces the heat necessary to promote further decomposition and leads to burning.

When burning is established, an overall (but over-simplified) equation can be written as shown in reaction (2.8):

$$4KNO_3(s) + 7C(s) + S(s) \longrightarrow$$
$$3CO_2(g) + 3CO(g) + 2N_2(g) + K_2CO_3(s) + K_2S(s) \quad (2.8)$$

where (s) = solid and (g) = gas phase.

At this stage, the spread of combustion from grain to grain is by a hot spray of molten potassium salts.

The important point to note here is that nearly all of the pre-ignition steps, including two endothermic reactions, take place in the vapour phase. Thus, if ignition is to occur at low pressures, sufficient heat energy must be available for a sufficient time period:

(1) to volatilise the initial reactants;
(2) to build up a local concentration of sufficient density to react;
(3) to supply the heat needed to take the reactions to the point where they become exothermic.

THERMAL IGNITION OF BLACK POWDER

It will be appreciated from the previous paragraphs that the ignition of explosives such as black powder is not an instantaneous process. Classically, as the temperature of the black powder is raised, slow decomposition sets in according to the process:

$$\text{Black powder} \longrightarrow \text{gaseous products} + \text{solid products} + \text{heat}$$

The rate of decomposition can be followed (for example) by measuring the volume of gas produced as a function of charge temperature as shown in Figure 2.3.

The dashed lines in Figure 2.3 represent the ignition temperature range for black powder which, depending on the conditions of the experiment, typically falls between about 280 and 400 °C.

Alternatively, the decomposition can be followed isothermally (at a

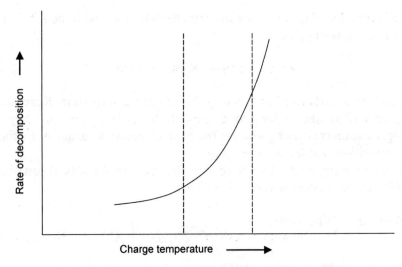

Figure 2.3   *The rate of decomposition of black powder as a function of charge temperature*

controlled temperature), again by measuring the rate of gas production. Figure 2.4 shows a typical result for isothermal decomposition.

The pre-ignition period begins with the application of the ignition stimulus, and ends with the start of a self-sustaining combustion. During this period, the rate of heat transfer to and the rate of heat production in the gunpowder become important in relation to the rate of heat loss from that portion of the material being ignited. As the temperature rises, chemical bonds break and molecular collisions occur while the number of molecules having sufficient energy (E) to react rise exponentially in accordance with the Arrhenius equation, (2.9),

$$k = Ae^{-E/RT} \qquad\qquad (2.9)$$

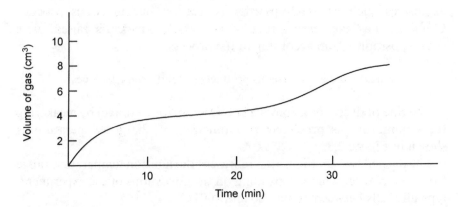

Figure 2.4   *Gas evolution from gunpowder measured at atmospheric pressure and 290 °C*

where k is the reaction rate constant, A is known as the frequency factor and is a constant related to the material, E is the activation energy in kJ mol^{-1}, R is the universal gas constant (8.314 J K^{-1} mol^{-1}) and T is the absolute temperature.

Since E is the numerator of a negative exponent, the reaction rate increases as E decreases.

The time to ignition can be expressed by equation (2.10) which is similar in form to the Arrhenius equation,

$$t = Ae^{-E/RT} \qquad (2.10)$$

where t is the time to ignition (*i.e.* ignition delay) at a temperature T in degrees absolute, and the other parameters are as described for equation (2.9).

Hence, by plotting the ignition delay (t) against temperature (T) a curve is obtained as in Figure 2.5.

Activation energies for black powders (and many other energetic materials) can be determined empirically from the logarithmic form of the Arrhenius equation (2.11):

$$\log t = -E/2.303RT + \log A \qquad (2.11)$$

By plotting $\log t$ *versus* $1/T$ a straight line is obtained whose slope is related to the activation energy by $E = [2.303(\text{slope})]R$ as shown in Figure 2.6.

Depending on the sulfur content of the black powder, activation energies ranging between 56 and 130 kJ mol^{-1} have been obtained, the

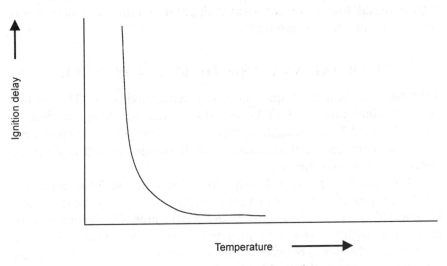

Figure 2.5   *Curve showing the variation of ignition delay with temperature*

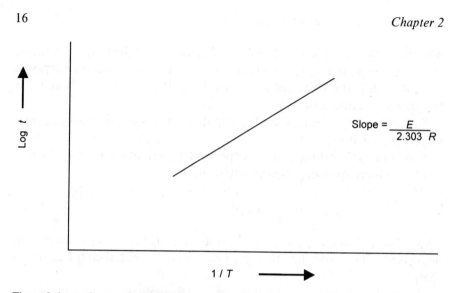

Figure 2.6 *Arrhenius plot for black powder*

higher figure being associated with sulfurless fine grain gunpowder (SFG).

Other work has shown that high volatile content charcoals lower the activation energy and reduce the ignition temperature of the black powder, while removal of volatile matter from the charcoal increases both activation energy and ignition temperature.

In summary, the thermal decomposition of black powder can be said to consist essentially of an initial reaction in which sulfur reacts with KNO_3 and/or volatile substances originating from charcoal followed by a main reaction involving charcoal and KNO_3.

High activation energies (greater than 300 kJ mol^{-1}) are associated with material that possesses greater thermal stability and that is less sensitive to an ignition stimulus.

THERMAL ANALYSIS OF BLACK POWDER

Thermal analytical techniques such as thermogravimetry (TG), differential thermal analysis (DTA) and differential scanning calorimetry (DSC) have all been successfully employed in studying the pyrotechnic reactions of energetic materials such as black powder, as well as of binary mixtures of the constituents.

TG is a method associated with mass change in which the mass of a substance is measured as a function of temperature whilst the substance is subjected to a controlled temperature programme. On the other hand, DTA is a method associated with temperature change in which the temperature difference between a substance and a reference material are subjected to a controlled temperature programme.

Multiple techniques are also possible such as TG–DTA whereby the mass loss of a sample and the DTA curve may be obtained simultaneously.

DSC is a method associated with enthalpy (heat content) change in which the difference in energy inputs into a substance and a reference material are subjected to a controlled temperature programme.

Analysis by TG

Typical thermogravimetric curves for black powder and its ingredients have been obtained as shown in Figure 2.7, and the main events depicted by the curves can be presented as in Table 2.1.

The TG curves in Figure 2.7 were obtained by heating the samples in a furnace at a rate of $15\,°C\ min^{-1}$ under an atmosphere of air or argon, and plotting the mass loss as a function of furnace temperature.

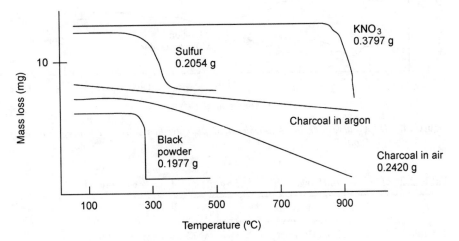

Figure 2.7 *Thermogravimetric curves for black powder and its ingredients*

Table 2.1 *Main thermal events depicted by the TG curves of Figure 2.7*

	Mass loss		
Substance	*From (°C)*	*To (°C)*	*Thermal events*
KNO_3	700	950	Thermal decomposition to potassium oxide
Sulfur	225	350	Sulfur vaporisation
Charcoal (heated in argon)	100	900	Loss of volatile constituents
Charcoal (heated in air)	50	950	Loss of moisture and oxyhydrocarbons then complete oxidisation
Black powder	250	275	Violent decomposition

Analysis by DTA

By contrast, DTA curves for black powder and its ingredients have been obtained as shown in Figure 2.8. The DTA curves are more complex than those obtained by TG and include crystal phase transitions as described in Table 2.2.

Apart from the additional information on heat release and phase changes given by DTA, the apparent discrepancy in the black powder ignition temperatures as determined by DTA and TG is explained by the fact that in DTA, the curves are plotted as a function of sample

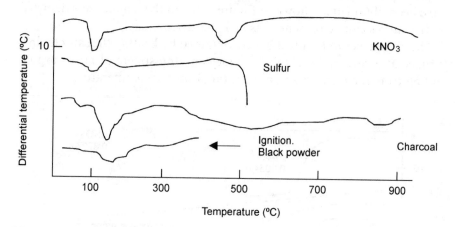

Figure 2.8 *DTA curves for black powder and its ingredients from a heating rate of 15°C min⁻¹*

Table 2.2 *Main thermal events depicted by the DTA curves of Figure 2.8*

| Substance | Mass loss | | Thermal events |
	From (°C)	To (°C)	
Black powder	95	200	Endotherm – overlapping peaks from crystalline transition of KNO_3, transition and fusion of sulfur and vaporisation of volatile matter from charcoal.
	250	300	Exotherm – pre-ignition followed by violent decomposition.
KNO_3	128	200	Endotherm – rhombic–trigonal crystal transition.
	334	360	Endotherm – KNO_3 melting.
Sulfur	105	110	Endotherm – rhombic–monoclinic crystal transition.
	120	125	Endotherm – sulfur melting.
	444		Endotherm – sulfur boiling.
Charcoal	100	300	Endotherm – loss of volatile constituents.

temperature, while with TG the curves are plotted as a function of the furnace temperature. If the sample in the TG thermobalance undergoes an exothermic reaction, its temperature will increase at a faster rate than that of the furnace temperature, while during an endothermic reaction the sample temperature will lag behind that of the furnace.

Analysis by DSC

DSC curves are represented in much the same form as DTA curves except that the integral of (the area beneath) the resulting peak is calculated by computer and is equivalent to the internal heat energy (enthalpy) change. In fact, the instrument has two separate control loops – one for average temperature control and the other for differential temperature control. The DSC response is based on the amount of power needed to correct any difference in temperature between the reference and the sample.

A typical DSC response for black powder is shown in Figure 2.9.

As with the DTA curves, the endotherms originate from phase transitions and volatilisation, while the single exotherm represents the violent decomposition of black powder.

Collectively, the thermal analysis techniques can be used to compare different batches of gunpowder and its constituents or to make more fundamental studies of, for example, the stability of the explosive under various physical or chemical conditions.

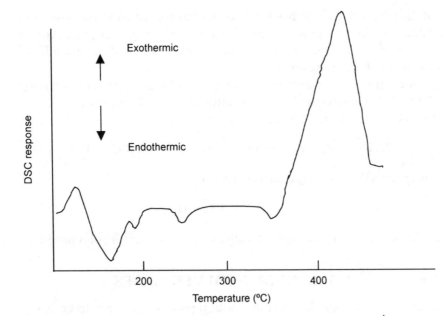

Figure 2.9 *DSC curves for black powder from a heating rate of 80°C min^{-1} in nitrogen*

As well as calculation of the temperatures and heat releases of the relevant exotherms, DSC data can be fitted to a differential equation (2.12) of the form,

$$\frac{df}{dt} = k(1-f)^n \qquad (2.12)$$

where f is the fractional degree of conversion of the sample to products (*i.e.* $f = 0$ at the start and 1.0 at the end of the exotherm), t is the time in seconds, the exponent n is a constant and k is the reaction rate constant (in units of s^{-1}). The rate constant varies during the course of the exotherm because of its strong dependence on temperature as shown earlier by the Arrhenius equation (2.9).

Such kinetic information can be used to estimate the thermal runaway times for the exotherm. This is the time taken for the explosive (under adiabatic conditions and considering a particular exotherm) to react at various initial temperatures.

The analysis of the kinetics of reacting solids is an area fraught with problems. This is because chemical kinetic theory strictly applies only to reactions of gases and liquids, and therefore absolute values of kinetic parameters derived from DSC must be treated with caution. However, it is quite reasonable to use the data in a comparative manner where this is derived from similar systems studied under the same conditions.

STOICHIOMETRY

The thermal analysis of black powders used for different purposes and containing differing proportions of ingredients can give an indication of the system stoichiometry by virtue of the rate and magnitude of measured exothermic events.

For example, the stoichiometry of a typical black powder containing KNO_3 (77%), charcoal (17%) and sulfur (6%) can be represented by the simplified equation (2.8), seen earlier as:

$$4KNO_3 + 7C + S \longrightarrow 3CO_2 + 3CO + 2N_2 + K_2CO_3 + K_2S \qquad (2.8)$$

This ignores the water gas reaction (2.13)

$$C + H_2O \rightarrow CO + H_2 \qquad (2.13)$$

and the presence of trace species such as water and minor solid products.

VOLUME OF EVOLVED GASES

From equation (2.8), the gaseous reaction products are seen to be $3CO_2$, $3CO$ and $2N_2$. Therefore, the reaction produces 8 moles of product gases from 4 moles of KNO_3 (404 g) plus 7 moles of charcoal (84 g) plus 1 mole of sulfur (32 g) which is equivalent to a total of 520 g of black powder.

To calculate the number of moles of gas produced per gram of black powder burned,

$$\frac{8}{520} = 0.015 \text{ mol g}^{-1}$$

and the volume of gas produced at standard temperature and pressure (STP) per gram of black powder burned is given by the Ideal Gas Law, equation (2.14).

$$V = \frac{nRT}{P} \tag{2.14}$$

STP refers to a temperature of 25 °C (298 K) and a pressure of 1 atmosphere (10^5 Nm^{-2}). Therefore,

$$V = \frac{0.015 \times 8.314 \times 298}{10^5} = 3.8 \times 10^{-4} \text{ m}^3 \text{ g}^{-1} \tag{2.15}$$

where R is the universal gas constant in units of J K^{-1} mol^{-1}, n is the number of moles, T is the absolute temperature and 1 J = 1 Nm.

If the actual temperature of the product gases is 2000 °C, the calculated volume at this temperature and one atmosphere pressure would be considerably greater, as shown by equation (2.16).

$$V_{2000\,°C} = \frac{2273}{298}(3.8 \times 10^{-4}) = 2.91 \times 10^{-3} \text{ m}^3 \text{ g}^{-1} \tag{2.16}$$

HEAT OF REACTION

The heat associated with the combustion of an energetic material such as black powder is called the heat of reaction. The heat of explosion is the heat of reaction associated with the rapid decomposition of such a material in an inert atmosphere.

Assuming the reaction of black powder given in equation (2.8) is at constant pressure; if only pressure–volume work is considered, the enthalpy change for the reaction at the temperature concerned is equal to the sum of the enthalpies of the products minus the sum of the enthalpies of the reactants [equation (2.17)],

$$\Delta H_{(reaction)} = \Sigma H_{(products)} - \Sigma H_{(reactants)} = Q_p \tag{2.17}$$

where Q_p is the heat effect.

Table 2.3 *Molar internal enthalpies of reaction products* $\bar{c}_p(T - T_o)$; $T_o = 25\,°C \approx 300$ K

Temp. (K)	CO kcal mol⁻¹	CO kJ mol⁻¹	CO₂ kcal mol⁻¹	CO₂ kJ mol⁻¹	H₂O kcal mol⁻¹	H₂O kJ mol⁻¹	H₂ kcal mol⁻¹	H₂ kJ mol⁻¹	O₂ kcal mol⁻¹	O₂ kJ mol⁻¹	N₂ kcal mol⁻¹	N₂ kJ mol⁻¹	NO kcal mol⁻¹	NO kJ mol⁻¹
700	4.99	20.89	7.68	32.16	5.98	25.04	4.75	19.89	5.22	21.85	4.93	20.64	5.11	21.39
900	6.60	27.63	10.33	43.25	8.00	33.49	6.21	26.00	6.91	28.93	6.52	27.29	6.75	28.30
1100	8.24	34.50	13.04	54.59	10.12	42.37	7.71	32.28	8.62	36.09	8.14	34.00	8.43	35.29
1200	9.08	38.01	14.43	60.41	11.22	46.98	8.47	35.46	9.49	39.73	8.97	37.55	9.28	38.86
1300	9.92	41.53	15.84	66.32	12.34	51.67	9.25	38.73	10.36	43.38	9.81	41.07	10.14	42.46
1400	10.78	45.14	17.25	72.23	13.48	56.44	10.04	42.03	11.25	47.10	10.65	44.59	11.01	46.10
1500	11.62	48.66	18.70	78.30	14.45	60.50	11.06	46.31	12.54	52.50	11.50	48.15	11.95	50.03
1600	12.47	52.21	20.07	84.03	15.56	66.15	11.77	49.28	13.30	55.69	12.33	51.63	12.79	53.55
1700	13.32	55.77	21.46	89.86	16.71	69.96	12.50	52.33	14.10	59.04	13.17	55.14	13.64	57.11
1800	14.16	59.29	22.85	95.67	17.87	74.82	13.26	55.52	14.91	62.43	14.00	58.62	14.49	60.67
1900	15.03	62.93	24.28	101.66	19.06	79.80	14.04	58.79	15.76	65.99	14.87	62.26	15.36	64.31
2000	15.81	66.20	25.73	107.73	20.29	84.91	14.84	62.14	16.63	69.63	15.73	65.86	16.23	67.96
2100	16.77	70.22	27.19	113.84	21.51	90.06	15.66	65.57	17.51	73.31	16.60	69.50	17.11	71.64
2200	17.65	73.90	28.66	120.00	22.79	95.42	16.49	69.04	18.41	77.08	17.47	73.15	17.99	75.32
2300	18.54	77.63	30.16	126.28	24.01	100.53	17.33	72.56	19.32	80.89	18.35	76.83	18.88	79.05
2400	19.42	81.31	31.62	132.39	25.28	105.85	18.19	76.16	20.24	84.74	19.23	80.52	19.78	82.82
2500	20.30	85.00	33.11	138.63	26.54	111.12	19.04	79.72	21.16	88.60	20.11	84.20	20.66	86.50
2600	21.19	88.72	34.61	144.91	27.83	116.52	19.91	83.36	22.11	92.60	20.99	87.90	21.56	90.27
2700	22.09	92.49	36.12	151.23	29.12	121.93	20.78	87.01	23.06	96.55	21.88	91.61	22.46	94.04
2800	22.98	96.22	37.63	157.56	30.42	127.37	21.67	90.73	24.01	100.53	22.77	96.34	23.36	97.81
2900	23.88	99.99	39.15	163.92	31.73	132.85	22.55	94.42	24.97	104.55	23.67	99.11	24.27	101.62

3000	24.78	103.75	40.68	170.33	33.04	138.84	23.48	98.31	25.94	108.61	24.56	102.83	25.17	105.39
3100	25.68	107.52	42.21	176.73	34.36	143.87	24.34	101.91	26.91	112.67	25.46	106.60	26.08	109.20
3200	26.57	111.25	43.73	183.10	35.67	149.35	25.21	105.64	27.88	116.73	26.35	110.33	26.98	112.97
3300	27.47	115.02	45.26	189.50	37.00	154.92	16.14	109.45	28.87	120.88	27.24	114.05	27.89	116.78
3400	28.37	118.79	46.80	195.95	38.33	160.49	27.05	113.26	29.85	124.98	28.15	117.86	28.81	120.63
3500	29.28	122.60	48.34	202.40	39.66	166.06	27.96	117.70	30.84	129.13	29.05	121.63	29.72	124.44
3600	30.18	126.36	49.88	208.85	40.99	171.63	28.88	120.92	31.83	133.27	29.95	125.40	30.63	128.25
3700	31.09	130.17	51.43	215.34	42.33	177.24	29.79	124.73	32.83	137.46	30.85	129.17	31.55	132.10
3800	32.00	133.98	52.97	221.79	43.67	182.85	30.71	128.58	33.83	141.65	31.76	132.98	32.47	135.95
3900	32.89	137.71	54.51	228.23	45.01	188.46	31.69	132.69	34.82	145.79	32.65	136.71	33.37	139.72
4000	33.80	141.52	56.06	234.72	46.35	194.07	32.55	136.29	35.82	149.98	33.56	140.52	34.29	143.57
4100	34.71	145.33	57.61	241.21	47.70	199.72	33.48	140.18	36.83	154.21	34.46	144.28	35.21	147.42
4200	35.62	149.14	59.17	247.74	49.05	205.37	34.41	144.07	37.83	158.39	35.37	148.09	36.13	151.28
4300	36.53	152.95	60.72	254.23	50.40	211.02	35.34	147.97	38.85	162.66	36.28	151.90	37.05	155.13
4400	37.44	156.76	62.28	260.77	51.76	216.72	36.27	151.86	39.85	166.85	37.19	155.71	37.97	158.98
4500	38.35	160.57	63.84	267.30	53.11	222.37	37.20	155.76	40.78	170.75	38.10	159.52	38.89	162.83
4600	39.26	164.38	65.40	273.83	54.47	228.07	38.13	159.65	41.88	175.35	39.00	163.29	39.81	166.68
4700	40.17	168.19	66.96	280.36	55.84	233.80	39.07	163.59	42.89	179.58	39.91	167.10	40.73	170.54
4800	41.08	172.00	68.51	286.85	57.18	239.41	40.00	167.48	43.90	183.61	40.81	170.87	41.64	174.35
4900	41.99	175.81	70.07	293.38	58.54	245.11	40.93	171.37	44.92	188.08	41.72	174.68	42.57	178.24
5000	42.90	179.62	71.64	299.96	59.90	250.80	41.87	175.31	45.94	192.35	42.63	178.49	43.49	182.09

If the reaction is a standard state reaction where the starting materials in their standard states react to give products in their standard states, and the standard heats of formation ($\Delta H^{\ominus}_{f,T}$) of the elements are assumed to be zero at any given temperature, then the standard heat of reaction, $\Delta H^{\ominus}_{(reaction)}$, is expressed as in equation (2.18):

$$\Delta H^{\ominus}_{T}(reaction) = \Delta H^{\ominus}_{f,T_0}(products) - \Delta H^{\ominus}_{f,T_0}(reactants) \qquad (2.18)$$

For example, the enthalpy of formation of CO_2 from combustion of the black powder constituent, charcoal, is given by reaction (2.19):

$$C(s) + O_2(g) \longrightarrow CO_2(g)$$

$$\Delta H^{\ominus}_{f,298} = -392.8 \text{ kJ mol}^{-1} \qquad (2.19)$$

Normally, the standard state is the most stable state at one atmosphere pressure and at the given temperature. Most tabular data, as used for the calculation of reaction temperatures, are given at 0 °C or 298 K. The overall calculation for the heat of reaction of black powder at different temperatures is simplified by using tabulated data of the enthalpy function, $H^{\ominus}_{T} - H_{298}$ for the reaction products, since no enthalpy measurements can be made in the sense of an absolute quantity.

Table 2.3 lists the molar internal enthalpies of black powder reaction products such as CO_2 where c_p values are the molar heat capacities of the products at constant pressure. Using these, it is possible to estimate the heat of reaction at a particular temperature by assuming two temperature values and summing up the internal enthalpies for the reaction products multiplied by their corresponding number of moles as in Table 2.4.

Plotting the sum of the enthalpies at the two reference temperatures yields an estimate for the heat of reaction for the 77:17:6 black powder composition over an extrapolated temperature range (see Figure 2.10).

Table 2.4 *Calculation of the sum of the enthalpies of the reaction products from gunpowder [reaction (2.8)] at two reference temperatures (2000 and 3000 K)*

Species	'n'	$H^{\ominus}_{3000} - H^{\ominus}_{298}$ (kJ mol^{-1})	kJ mol^{-1} × 'n'	$H^{\ominus}_{2000} - H^{\ominus}_{298}$ (kJ mol^{-1})	kJ mol^{-1} × 'n'
K_2CO_3	1	480[a]	480	344.38	344.38
N_2	2	102.83	205.66	65.86	131.72
CO	3	103.75	311.25	66.20	198.60
CO_2	3	170.33	510.99	107.73	323.19
K_2S	1	160[a]	160	140[a]	140
			$\Sigma = 1667.90$		$\Sigma = 1137.89$

[a] Estimated.

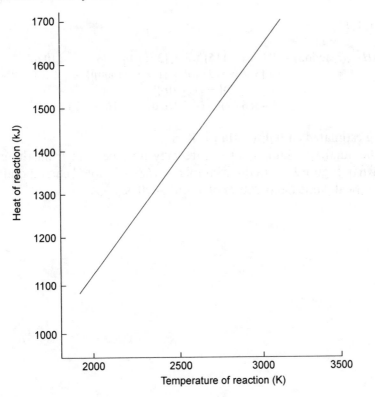

Figure 2.10 *Estimate of the heat of reaction for the 77:17:6 black powder composition from the sum of the product enthalpies at two reference temperatures*

For a reaction temperature of, say, 2500 K, the theoretical heat of reaction is approximately 1410 kJ mol^{-1}.

TEMPERATURE OF REACTION

Further use of tabulated data (such as those in Table 2.3) enables an estimate to be made of the temperature of the reaction of black powder. Using equation (2.18), the standard enthalpy change may be calculated from the standard heats of formation of the reactants and products as in equation (2.20).

$$\Delta H_{298}^{\ominus}(\text{reaction}) = \Delta H_{f,298}^{\ominus}(\text{products}) - \Delta H_{f,298}^{\ominus}(\text{reactants}) \qquad (2.18)$$

Thus,

$$\begin{aligned}
\Delta H_{298}^{\ominus}(\text{reaction}) = \; & \{1[\Delta H_f^{\ominus}(K_2CO_3)] + 2[\Delta H_f^{\ominus}(N_2)] \\
& + 3[\Delta H_f^{\ominus}(CO)] + 3[\Delta H_f^{\ominus}(CO_2)] + 1[\Delta H_f^{\ominus}(K_2S)]\} \\
& - \{4[\Delta H_f^{\ominus}(KNO_3)] + 7[\Delta H_f^{\ominus}(C)] + 1[\Delta H_f^{\ominus}(S)]\}
\end{aligned}$$
$$\qquad (2.20)$$

Therefore,

$$
\begin{aligned}
\Delta H^{\ominus}_{298}(\text{reaction}) &= \{[1 \times (-1151.0)] + [2 \times (0)] + [3 \times (-110.6)] \\
&\quad + [3 \times (-393.8)] + [1 \times (-1000)]\} - \{[4 \times (-494.0)] \\
&\quad + [7 \times (0)] + [1 \times (0)]\} \\
&= (-3664.2) - (-1976.0) = -1688.2 \text{ kJ mol}^{-1}
\end{aligned}
$$

using estimated enthalpy data for K_2S.

The adiabatic flame temperature may now be read from the graph shown in Figure 2.10. In this example, -1688 kJ mol^{-1} corresponds to a theoretical flame temperature of about 3070 K.

Chapter 3

Rockets

PROPELLANT

The initial development of the firework rocket and the military rocket probably occurred during the same period in history. Both used black powder as the rocket propellant.

In sending a rocket into the sky we are calling into action several laws of physics and chemistry, and the same laws apply whether the application is a small firework rocket weighing a few ounces or a solid propellant 'booster' for the space shuttle containing around 300 tons of propellant. These fundamental processes may be conveniently divided into internal ballistics and external ballistics.

Internal ballistics

When black powder is used to propel rockets it is classed as a composite propellant (where the fuel and oxidiser are intimately mixed) and forms part of a rocket motor in which the powder is compressed to form a monolithic single grain inside a combustion chamber as shown in Figure 3.1.

The importance of compressing the gunpowder grain is to control the rate of burning by ensuring that the surface of the grain is not porous to the hot combustion gases. Such penetration would result in a progressive increase in the rate of combustion and loss of ballistic control. Interestingly, the burning rate laws that were discovered for the burning of gunpowder in the 19th century are equally applicable to more modern solid propellants such as cordite.

Piobert's law of 1839 states that 'burning takes place by parallel layers where the surface of the grain regresses, layer by layer, normal to the surface at every point'. Thus the combustion gases flow in the opposite direction to the rate of combustion progress (or surface regression).

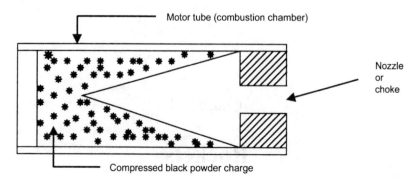

Figure 3.1 *The essential components of a firework rocket motor*

Vieille's law of 1893 is mathematical, but if a book on fireworks was permitted to contain just one equation, this would be it [equation (3.1)]:

$$R_B = bp^n \qquad (3.1)$$

The above relationship basically shows that the rate of burning (R_B) normal to the burning surface is dependent on the ambient pressure (p). Although an over-simplification, it is universally accepted that the higher the pressure, the greater the heat transfer onto the surface of the grain and hence, the higher the rate of burning.

The constants 'b' and 'n' are dependent upon the chemical composition of the grains and their initial temperature. The burning rate equation (3.1) is based on the various empirical measurements and differs with the type of propellant. For one type of gunpowder the equation might be expressed as follows:

$$R_B = 3.38p^{0.325} \qquad (3.2)$$

In this example, R_B is the linear rate of burning in mm s^{-1} and p is the rocket motor chamber pressure in pounds per square inch (psi).

Substituting a pressure of, say, 300 psi into equation (3.2) gives a rate of burning of 21.6 mm s^{-1}, whereas a pressure of 1000 psi (6.9×10^6 mN m^{-2}) results in a rate of 31.9 mm s^{-1}. These values may be plotted on log–log graph paper to give a straight line as in Figure 3.2, whose slope is equal to the pressure exponent, n, whose value is 0.325, from equation (3.3):

$$n = \frac{\log_{10}(R_B) - \log_{10}(R_B - 1)}{\log_{10}(p) - \log_{10}(p - 1)} \qquad (3.3)$$

When designing a gunpowder charge for a firework rocket we can use Vieille's law to determine the burning speed of the charge at a particular

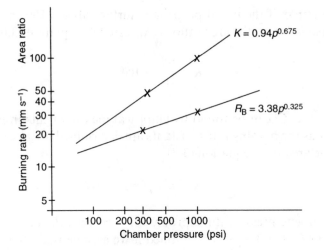

Figure 3.2 *Burning rate and area ratio data from black powder propellant*

pressure, and a similar law to calculate the pressure operating within the rocket body or 'motor'. But in order to estimate the operating pressure we need to know two further things: the area of the burning propellant surface and the cross-sectional area of the nozzle (or choke) through which the powder gases pass. You will notice on firing a rocket that there is an initial rapid acceleration, after which the flame dies down and the rocket 'coasts'. The sudden acceleration is because the area of propellant is increased near the nozzle. In fact, the gunpowder charge is pressed with a spike so that there is a deep 'cone' at the ignition end, this serving to increase the surface area of the propellant to perhaps 100 times that of the area of the nozzle. As propellant is consumed, its area diminishes and the gas flow or 'thrust' reduces.

The ratio of propellant area (A_p) to nozzle area (A_N) is known as the propellant area ratio, K, *i.e.*

$$K = \frac{A_p}{A_N} \tag{3.4}$$

In fact, an area ratio equation such as (3.4) can be compared with the burning rate equation (3.2) by taking note of the fact that 'n' [or 0.325 in equation (3.2)] becomes '$1 - n$' (or 0.675) in the area ratio equation. Therefore, equation (3.4) may be written,

$$K = Cp^{1-n} \tag{3.5}$$

where C is a further constant dependent upon the propellant.

For example, if the initial propellant surface area is 100 cm^2 and the nozzle area is 1 cm^2, the area ratio is expressed by equation (3.6):

$$K = \frac{100}{1} = 100 \qquad\qquad (3.6)$$

The constant, C, can be found by empirical measurements in much the same way as the burning rate constants are determined, to give a value of 0.94 in the area ratio equation (3.7):

$$K = 0.94p^{0.675} \qquad\qquad (3.7)$$

The complete pressure–time profile for the firework rocket motor as depicted in Figure 3.1 can be obtained if we assume that the propellant surface decreases in area during burning inwardly in parallel layers (according to Piobert's law).

Typical results from applying the above concepts are presented in Table 3.1 where the chamber pressures and burn times relate to burning in increments to give an overall burn time in seconds.

The data in Table 3.1 may then be plotted as in Figure 3.3 to give the pressure–time profile for the firework rocket motor.

Finally, the thrust (F) of a rocket motor is also known to be related to the rate of burning and is obtained *via* the mass flow rate using the relationship given in equation (3.8):

$$F(\text{th}) = \bar{R}.I_{sp}.D.\bar{A}_p \qquad\qquad (3.8)$$

In this equation $F(\text{th})$ is the theoretical thrust, $\bar{R}$ is the average rate of burning, I_{sp} is a performance parameter related to the propellant called the specific impulse, D is the propellant density and $\bar{A}_p$ is the average area of the burning surface.

Table 3.1 *Pressure–time data for firework rocket motor*

Inward burning increment (cm)	A_p (cm^2)	A_N (cm^2)	K	P (psi)	R_B (mm s^{-1})	Incremental burn time (s)	Overall burn time (s)
0.0	50	1	50	380	22	0.91	0.91
0.2	40	1	40	275	21	0.95	1.86
0.4	30	1	30	180	18	1.11	2.97
0.6	20	1	20	100	15	1.33	4.30
0.8	10	1	10	37	11	1.82	6.12

Burning rate equation, $R_B = 3.38p^{0.325}$. Area ratio equation, $K = 0.94p^{0.675}$.

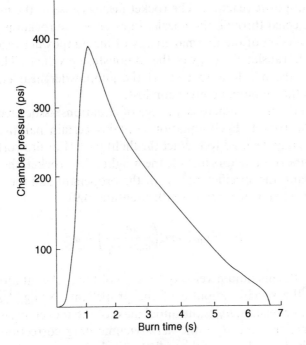

Figure 3.3 *Pressure–time profile of a rocket motor*

For the rocket motor described above, a typical theoretical thrust can be obtained from the following parameters:

$$\bar{R} = 20 \text{ mm s}^{-1} \equiv 0.020 \text{ m s}^{-1}$$
$$I_{sp} = 600 \text{ Ns kg}^{-1}$$
$$D = 1800 \text{ kg m}^{-3}$$
$$\bar{A}_p = 30 \text{ cm}^2 \equiv 0.0030 \text{ m}^2$$

Substituting the above values into equation (3.8) gives a theoretical thrust of 64.8 Newtons (N). One Newton is equivalent to 0.22 lb force and so the thrust from the rocket motor is about 14.3 lbs. For a firework rocket weighing 1 lb a thrust to weight ratio of 14.3 is perfectly reasonable and such a design would be cheap and not highly stressed.

External ballistics

In the propulsion of a solid propellant rocket, relatively small masses of materials are ejected through the nozzle at a very high velocity. We know from Newton's Third Law of Motion that to every action there is an equal and opposite reaction. The rocket flies because of the momentum of matter ejected through the nozzle. In essence, the rocket is propelled by the conversion of the thermal energy of the gunpowder reaction into the directed, translated energy of the combustion products. The primary function of the nozzle is to convert the gunpowder heat energy into thrust with the minimum conversion loss.

In external ballistics there is a range of equations associated with the flight of the rocket. Each equation contains certain parameters that, collectively, may be used to predict the flight path. For firework rockets, the most important factors include the weight of the rocket, the weight of the propellant, the specific impulse of the propellant and the shape and size of the rocket in accordance with equation (3.9),

$$\Delta v = I_{sp}.g.\log_e\left(\frac{m_o}{m_o - m_p}\right) - d_c \qquad (3.9)$$

where Δv is the maximum velocity (m s^{-1}) of the rocket at motor burn-out, I_{sp} is the specific impulse of the propellant (Ns kg^{-1}), g is the gravitational constant, m_o is the initial mass of the rocket, m_p is the mass of the propellant, and d_c is the aerodynamic drag correction which is related to the frontal area of the rocket, the air density and to the square of the rocket velocity.

The specific impulse is a figure of merit for a particular propellant and is defined as the thrust that can be obtained from an equivalent rocket which has a propellant flow rate of unity.

In summary, the rocket motor acts as a gas generator whose force is sufficient to overcome the combined effects of gravity and the drag which act in a direction opposite to the flight path. Stability in flight is crudely provided by the stick or tail that ensures the centre of gravity is forward of the centre of pressure. At the centre of pressure the combined effects of cross-wind force (perpendicular to the direction of motion) and the drag force act to restore the tail-stabilised rocket to alignment with the motion of the centre of gravity. Look at a firework rocket carefully and you will see it 'shuttlecock' into the wind, by virtue of the tail (Figure 3.4).

ROCKET DESIGN AND MANUFACTURE

Firework rockets vary in size (and price) from small number 4 rockets which will contain a thunderflash or a coloured star, to number 6 with more coloured stars, then to numbers 7 and 8 with larger star heads, to number 10 which provides the largest bouquet of stars. There are screechers, whistlers and maroon-type rockets with exploding heads, all in various calibres.

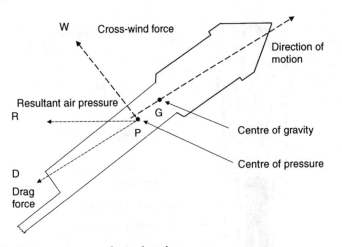

Figure 3.4 *Forces acting on a firework rocket*

For the more adventurous, parachute rockets are available in 22 and 25 mm calibre, or even 'festoons' or bag rockets with sticks up to three metres in length. The latter are definitely fired individually!

In manufacture, fuel-rich gunpowder is pressed in increments into the rocket case (or motor) using a long tapering spigot, which, upon withdrawal, leaves a charge with a central hole or conduit tapering outwards towards the nozzle (or choke). This provides a surface area of propellant necessary for the rapid generation of thrust on ignition as described previously. The case is either constricted or a clay nozzle is pressed at the end of the motor, which is then closed off with a length of quickmatch through the nozzle and an outer wrapping of blue touch-paper applied.

When a thunderflash effect is required, a square of touch-paper is placed on the rear propellant surface inside the case at the opposite end to the choke, and a 'hairpin' of quickmatch is placed inside the case in contact with the square of touch-paper. Some flash composition, based on potassium perchlorate, barium nitrate and aluminium, is then poured around the quickmatch and the head of the case is closed off with a pressing of clay or by simply being taped over. The all-important stick is then glued to the outside of the case using the wrappings of the paper label for support.

With larger payloads, such as those containing stars, a plastic cone is normally employed and is designed to be integral with the flight tube. The rocket motor can be pre-pressed into a cardboard motor tube and the whole thing inserted as a cartridge grain before being choked and fused in the normal way.

Inside the plastic cone at the head of the rocket might be a payload of small green stars, each about the size of a pea, containing ingredients such as barium chlorate, potassium chlorate, gums and binders.

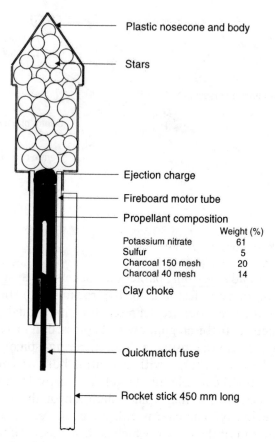

Figure 3.5   *Bouquet rocket*

A length of match or a pressing of delay fuse connects the rocket motor with the payload so that when the motor burns-out the fuse is lit and communicates with an ejection charge of gunpowder which ignites and ejects the stars.

A schematic diagram of a 35 mm bouquet rocket is presented in Figure 3.5.

Parachute rockets work on much the same principle, except that the ejection charge pushes out a tube of flare composition, or any other composition, which is connected to the parachute which itself is folded beneath the end-cap at the top of the rocket.

RECENT DEVELOPMENTS

Obviously, solid rocket propellant compositions have advanced considerably since the days of gunpowder, and it might be mentioned, in passing, that even as late as the Second World War, the Allies held the deluded belief that the large rockets being developed by the Germans were based on solid propellant. However, if one calculates the amount of solid propellant required to lift a payload consisting of one ton of high explosive plus all of the hardware associated with the rocket, one arrives at an impossibly high all-up weight. It was therefore (incorrectly) argued that the V2 must either have a trivial payload or that the range must be insufficient for the weapon to reach London.

The firework mentality also decreed that even if the range was achievable, in order for an unguided rocket to follow a rainbow trajectory across the English Channel and fall, under gravity, onto London, the launch angle would have to be impossibly precise – about half a degree from the vertical, in fact.

Both arguments were flawed, of course, and Allied Intelligence later showed that the 14 ton V2 rocket bomb was in fact powered with liquid propellant, was guided using gyroscopes, and did indeed carry one ton of explosive. And on the 3rd September 1944 the first V2 fell on London. Some 'firework'!

There is nothing to prevent the firework manufacturers from using military-type propellants, of course, and this is exactly what the Russians do when making rockets to reach astonishing heights in firework displays above the tall buildings in Moscow.

Military propellants are based on relatively powerful oxidisers and fuels of high calorific value in order to develop an improved thrust or impulse. Thus the most commonly-used oxidisers are potassium per-

Table 3.2 *Examples of ingredients used for composite propellants*

Oxidiser	Formula
Ammonium perchlorate	NH_4ClO_4
Ammonium nitrate	NH_4NO_3
Potassium perchlorate	$KClO_4$
Nitronium perchlorate	NO_2ClO_4
Hydrazinium nitroformate	$NH_2NHC(NO_2)_3$
Binder	
Polysulfide	$(C_2H_4S_4)_n$
Poly(vinyl chloride)	$(C_2H_3Cl)_n$
Polyurethane $[O.(CH_2)_4OOCNH.(CH_2)_6NHCOO.(CH_2)_4O]$	$(C_{16}H_{30}N_2O_6)_n$
Curing or bonding agents	
Toluene-2,4-diisocyanate	$C_9H_6N_2O_2$
Triethanolamine $N(CH_2CH_2OH)_3$	$C_6H_{15}NO_3$
Metal fuel	
Aluminium	Al
Magnesium	Mg
Beryllium	Be
Boron	B

chlorate, ammonium perchlorate or more esoteric compounds such as hydrazinium nitroformate. Metallic fuels include aluminium, magnesium and beryllium, while binders are mainly hydrocarbons such as polybutadiene, polyisobutylene, polyurethane or poly(vinyl chloride) (PVC) as presented in Table 3.2.

If smokelessness is required, then a double-base propellant such as cordite (nitrocellulose/nitroglycerine) can be used. Most of the military propellants enjoy about three times the specific impulse of gunpowder and are deliberately formulated to be fuel-rich so that the exhaust tends to contain carbon monoxide in preference to carbon dioxide, which is a heavier gas. Lighter gases are preferred because they can be accelerated to higher velocities through the nozzle and this factor contributes to a higher value of specific impulse. On the other hand, solid propellants based on gunpowder produce carbon dioxide as well as large quantities of smoke, neither of which are conducive to high values of thrust or impulse.

But who wants to see a supersonic rocket full of smokeless propellant disappear into the heavens like a flash of light when one can witness the leisurely ascent of a firework rocket as it climbs leaving a graceful trail of sparks?

Chapter 4

Mines and Shells

CALIBRES

Without doubt, mines and shells are the most important of the display fireworks in the operator's arsenal: shells give the well known spectacular aerial effects, while mines operate from ground level often as a supplement to the shells.

Although the most common calibres of the mortar tubes associated with these fireworks are 50, 75 and 100 mm, the size of shells seems to know no limit – the record at the moment stands at a calibre in excess of 1 metre or 1000 mm. Obviously, where such astonishingly large devices are concerned, the mortar tubes are no longer made from simple overlapping spirals of Kraft paper but preferably of stainless steel!

CONSTRUCTION OF SHELLS

The shell itself is a clever arrangement of stars built inside a spherical or cylindrical cavity, at the centre of which is a bursting charge of gunpowder or similarly forceful powder. A further quantity of gunpowder, held in a paper pouch, is located outside the base of the shell, and this acts as the lifting charge when ignited by a length of piped match that projects out of the pouch to run from the bottom of the mortar tube, where the shell lies, to the top of the tube where it can be lit using a fuse extension. When the match is lit, the gunpowder in the pouch is caused to explode, thereby driving the shell out of the mortar tube and up into the sky. Figure 4.1 shows the construction of a typical 75 mm shell.

A length of delay fuse connects to the inside of the shell and this fuse is lit from the primary explosion. After a delay of several seconds, and ideally when the shell has reached maximum height, the connecting fuse delivers its flame to the gunpowder bursting charge, which promptly explodes, igniting the ball of stars and driving them outwards, with

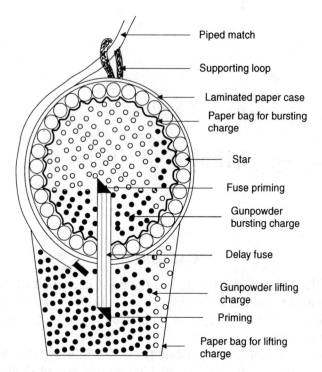

Piped match

Supporting loop

Laminated paper case

Paper bag for bursting charge

Star

Fuse priming

Gunpowder bursting charge

Delay fuse

Gunpowder lifting charge

Priming

Paper bag for lifting charge

Figure 4.1   *Schematic representation of a 75 mm chrysanthemum shell*

approximately equal force, in the form of a rapidly-expanding ball, a flash or any other desired effect.

The shell case can be made from paper, wood or similar material, reinforced with string. For a given calibre, the cylindrical shell holds more stars and is the more straightforward to produce, especially when multiple breaks are required. The much revered ten-break shell (yes, it does burst ten times in the sky!) comprises separate compartments, each connected with a delay fuse to give exploding effects from the apogee (top) of the flight, and then nine times afterwards while the shell is falling back to earth. Of course, multiple bursts can be achieved by using stars that explode during burning – a sort of star within a star – but that is not to detract from the skill and ingenuity and time involved in the making of complex fireworks such as star shells.

Early methods of construction of shells involved the fabrication of hollow spheres using paper strips pasted together in moulds. However, it was difficult to obtain an even burst in this way because, depending on where the joins were located in the shell case, the bursting shell was prone to form non-symmetrical fragments.

Modern methods of shell construction ensure an even distribution of stars on bursting by paying particular attention to the bursting charge in relation to the strength of the shell case. For example, a weak outer case can be adequately burst by a relatively small gunpowder charge. On the other hand, a strongly-made case must be burst by a charge that expands more slowly. Plastic has replaced paper in the manufacture of many round shells which, besides being waterproof, offers the advantage of unit construction whereby the lifting charge may also be contained inside the plastic case.

The most spectacular effects are achieved using cylinder shells in which cylindrical 'parcels' are filled with stars, whistles or hummers contained within separate compartments. The difficulty in producing cylinder shells arises from the many operations during manufacture, which include forming the cylinder from paper and cardboard discs, loading the stars, adding lifting charges and fuses, and closing the ends of the cylinder before wrapping the entire unit with string to provide an even burst before adding outer layers of Kraft paper to cover the string.

Internal Ballistics

In using gunpowder as a lifting charge for a shell, the propellant is being required to deliver its gas over a much shorter time interval than that employed in a rocket. In consequence, the geometrical form of the propellant is quite different, as is the propellant composition. The internal ballistics calculations call for the introduction of gun-type parameters that would be less appropriate in studying rocket motor ballistics. These include the propellant co-volume (the volume occupied by the combustion products), the geometrical form coefficient (the shape and size of the propellant grains) and the vivacity (quickness) of the propellant, together with the cross-sectional area of the mortar tube and the mass of the shell.

Table 4.1 gives examples of the properties of a typical lifting charge propellant. Note that the composition is no longer fuel-rich, but is balanced for 'quickness'.

Table 4.1 *Some gunpowder properties used in ballistics calculations*

Composition (%)	KNO$_3$ 74.3, charcoal 15.3, sulfur 9.4, moisture 1.0
Grain size (mm)	1.0–1.2
Force, $f = nrt$, (kgm kg^{-1})	3837
Vivacity, A (s^{-1})	4.17
Heat of explosion, Q (kJ)	1550
Explosion temperature, T (K)	2050
Evolved gas volume per kg at STP, V (dm^3)	404
Burning rate equation	$R_B = 2.5p^{0.5}$

Pressure in the Mortar Tube

The maximum pressure generated in the mortar tube has been shown to be related to the ballistic parameters presented in equation (4.1),

$$P_m = \frac{Af}{AG} \frac{1}{(1/\Delta) - \eta}(D_m) \qquad (4.1)$$

where P_m is the maximum pressure generated (kg dm^{-2}), A is the vivacity or quickness of the propellant (s^{-1}), Δ is the loading density of the propellant (kg dm^{-3}), f is the explosive force of the propellant (kgm kg^{-1}), G is the grain shape function, η is the propellant co-volume (dm^3) and D_m is the pressure function for maximum pressure.

Substituting typical values into equation (4.1) enables an estimate of the maximum mortar tube pressure to be made [equation (4.2)],

$$P_m = \frac{4.17 \times 3837}{4.17 \times 0.051} \times \frac{1}{(1/0.104) - 0.55}(0.212) \qquad (4.2)$$

from which $P_m = 1750$ kg dm^{-2}, or 17.5 kg cm^{-2}, or about 250 psi.

Burning Time of the Propellant

At a maximum generated pressure of 250 psi, the burning time of the propellant can be estimated from the linear burning rate, equation (4.3):

$$R_B = 2.5p^{0.5} \qquad (4.3)$$

Thus,

$$R_B = 2.5 \times 250^{0.5} \qquad (4.4)$$

and therefore

$$R_B = 39.5 \text{ mm s}^{-1}.$$

If the propellant grain size averages 1.1 mm, then the average radius (or web size) = 0.55 mm. At a rate of burning of 39.5 mm s^{-1}, the average burning time, T_B, will be given by equation (4.5):

$$T_B = \frac{\text{web size}}{R_B} = \frac{0.55}{39.5} = 0.014 \text{ s or } 14 \text{ ms} \qquad (4.5)$$

Therefore, the pressure–time curve for the combustion of the gunpowder lifting charge will be approximately as illustrated in Figure 4.2.

When ignited, the lifting charge causes the gas pressure behind the shell to rise and the rate of rise increases rapidly as the burning of the charge progresses. The maximum rate of rise of pressure is about

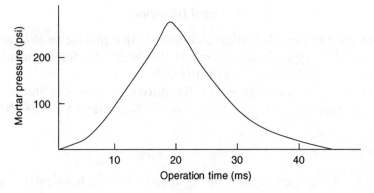

Figure 4.2 *Approximate pressure–time curve for combustion of lifting charge within a mortar tube*

30 000 psi s^{-1}. About 14 ms after initiation of the charge, the pressure reaches its maximum value when the rate of increase of pressure due to the burning of the charge is equal to the rate of pressure reduction due to the movement of the shell and 'leakage' of gases past the shell providing an increasing volume into which the gases can expand. After this, the pressure gradually decreases as the shell travels towards the muzzle of the mortar tube and then falls to zero after the shell leaves the tube.

Muzzle Velocity of the Shell

A further equation based on gun ballistics has been proposed that allows an estimation to be made of the muzzle velocity of a shell [equation (4.6)],

$$V = \frac{1}{AG} \times \frac{\sigma}{\mu}(W) \qquad (4.6)$$

where V is the estimated muzzle velocity (dm s^{-1}), σ is the cross-sectional area of the mortar tube (dm^2), μ is the assumed mass of the shell (kg), W is the velocity function and A and G have the definitions as listed for equation (4.1).

Substituting appropriate values into equation (4.6) gives the muzzle velocity of the shell as in equation (4.7):

$$V = \frac{1}{0.213} \times \frac{1.81}{0.0132} \times 1.97 = 1270 \text{ dm s}^{-1} \qquad (4.7)$$

or

$$V = 127 \text{ m s}^{-1}$$

External Ballistics

The simplest external ballistics calculation is that in which a shell is fired vertically. This type of calculation has also been extended to include the case where a 'dud' shell falls back to earth.

Equations (4.8)–(4.15) show how the altitude attained and time of flight can be estimated from basic parameters and Newtonian Laws of Motion.

Altitude Attained

First, the surface area (S) of the shell cross-section is calculated using equation (4.8),

$$S = dB^2/4 \qquad (4.8)$$

where dB is the diameter of the shell (dm). Hence,

$$S = 3.142(0.136)^2/4 = 0.0145 \text{ dm}^2.$$

Secondly, the ballistic parameter, a, is determined from equation (4.9),

$$a = \sqrt{pB/KS} \qquad (4.9)$$

where pB is the mass of the shell (kg) and K is the coefficient of air resistance (kg s^2 m^{-4}).

Substituting values into equation (4.9) enables the ballistic parameter to be calculated:

$$a = \sqrt{1.252/0.031 \times 0.0145} = 52.8 \text{ m s}^{-1} \qquad (4.10)$$

Thirdly, the altitude parameter, x, is calculated using equation (4.11),

$$x = V_1/a \qquad (4.11)$$

where V_1 is the muzzle velocity of the shell from equation (4.7) and a is the ballistic coefficient from equation (4.10). Therefore,

$$x = 127/52.8 = 2.40 \qquad (4.12)$$

We must now turn to ballistics tables such as Table 4.2 to determine the ratio of altitude:muzzle velocity squared.

From Table 4.2 it can be seen that a value of x of 2.40 corresponds to a H/V_1^2 ratio of 0.0169. Knowing the muzzle velocity of the shell (127 m s^{-1}) enables an estimate of the maximum altitude to be made as in equation (4.13):

$$H = 0.0169 \times V_1^2 = 0.0169 \times 127^2 = 273 \text{ m} \qquad (4.13)$$

Table 4.2 *The ratio of altitude : muzzle velocity squared for the altitude parameter,* x

x	H/V_1^2	x	H/V_1^2
0.20	0.0500	2.20	0.0186
0.40	0.0473	2.40	0.0169
0.60	0.0438	2.60	0.0155
0.80	0.0394	2.80	0.0142
1.00	0.0354	3.00	0.0131
1.20	0.0316	3.20	0.0121
1.40	0.0283	3.40	0.0112
1.60	0.0253	3.60	0.0104
1.80	0.0228	3.80	0.0097
2.00	0.0205	4.00	0.0090

Time of Flight

The time of flight (T) of the shell is determined from equation (4.14),

$$T = \left(\frac{a}{g}\right) \arctan \frac{H}{V_1^2} \tag{4.14}$$

where g is the acceleration due to gravity (dm s^{-2}) and a, H and V_1 are as defined previously. Hence:

$$T = \left(\frac{52.8}{9.80}\right) \arctan 0.0169 \tag{4.15}$$

or

$$T = 5.39 \times 0.97 = 5.2 \text{ s}$$

Descent of a 'Dud' Shell

Approximate estimates for the rate of descent if a shell fails to explode can be found by extending the data obtained in the previous section, *i.e.*:

$$H = 273 \text{ m and } a = 52.8 \text{ m s}^{-1}$$

Three equations (4.16)–(4.18) govern the rate of descent,

$$-H = \frac{2.303}{2g}|a|^2 \log(1 - z^2) \tag{4.16}$$

$$-V_2 = az \tag{4.17}$$

$$T_2 = \frac{2.303}{2g}|a| \log\left(\frac{1-z}{1+z}\right), \quad -1 < z < 0 \tag{4.18}$$

where H is the fall height (m), z is a ballistic parameter, V_2 is the falling velocity at the ground (m s^{-1}) and T_2 is the fall time (s).

Therefore, from equation (4.16),

$$\log(1 - z^2) = \frac{-2gH}{2.303a^2} = \frac{-2 \times 9.80 \times 273}{2.303 \times 52.8^2} = -0.833$$

$$(1 - z^2) = 0.147, \ i.e. \ z = -0.924$$

and from equation (4.17),

$$-V_2 = az = (-52.8) \times 0.924 = -48.8 \, \mathrm{m \, s^{-1}}$$

Finally, the fall time is estimated by applying equation (4.18),

$$T_2 = \frac{2.303}{2 \times 9.80}(52.8)\log\frac{1 + 0.924}{1 - 0.924} = 8.7 \, \mathrm{s} \qquad (4.19)$$

which tells us that if a 150 mm shell fails to explode, we have rather less than ten seconds in which to get out of the way!

Mortar Tubes

For display shells of 200 mm and above, reinforced fibreglass mortar tubes are invariably used. Welded iron tubes have long been considered to be amongst the safest because, in the event of a pressure-burst, the tube will split in the vicinity of the weld, thus creating a predictable danger zone. With seamless tubes the fragmentation is completely random and creates a shrapnel hazard for the fireworks operator.

Recent work in the USA has demonstrated that a high degree of safety can be achieved by using a tube within a tube, the concentric gap being filled with high-impact, expanded foam. Placing these tubes in a staggered formation in a metallic firing crate shows the assembly to be surprisingly resistant to damage, even when shells are deliberately exploded inside the tubes. The staggered arrangement prevents the explosion from communicating its effects with neighbouring tubes and shells.

Table 4.3 gives some estimates of the results to be expected from firing shells from 50, 75, 100 and 150 mm mortars.

In practice, a 120 mm shell can attain an altitude in excess of 200 metres in approximately five seconds and burst to give a spectacular effect stretching for perhaps 80 metres across the sky.

Table 4.3 *Estimates of shell performance*

	Shell size (mm)			
	50	75	100	150
Mortar fibreboard				
Inside diameter (mm)	52	76	104	155
Length (mm)	350	420	600	1050
L:D ratio	6.73	5.33	5.76	6.77
Wall thickness (mm)	5	10	10	15
Shell				
Diameter (mm)	46	70	94	142
Weight (kg)	0.065	0.210	0.390	0.950
Lift charge (black powder)				
Weight (kg)	0.004	0.013	0.025	0.075
Weight ratio (powder:shell)	0.061	0.062	0.064	0.079
Muzzle velocity (m s^{-1})	93	105	125	130
Maximum altitude (m)	90	110	130	275
Rise time (s)	3.7	4.0	4.1	5.4

Energy Transfer Efficiency

Finally, an estimate can be made of the amount of energy expended by the gunpowder lifting charge in aspects such as warming up the mortar tube and overcoming the friction of the shell.

Assuming that a 75 mm shell is fired by a 13 g lifting charge to give a muzzle velocity (V) of 105 m s^{-1} for a shell of mass (m) 0.21 kg, the kinetic energy (KE) of the shell is given by equation (4.20),

$$KE = \frac{mV^2}{2g} \qquad (4.20)$$

where g is the gravitational constant (9.8 m s^{-2}).

Therefore, the kinetic energy of the shell at the muzzle will be:

$$\frac{0.21 \times 105^2}{2 \times 9.8} = 118 \text{ kg m or } 1.16 \text{ kJ} \qquad (4.21)$$

The muzzle energy can be compared with the overall explosion energy from the gunpowder lifting charge, which is typically about 2000 kJ kg^{-1}, and hence the available energy from the 0.013 kg lifting charge is $2000 \times 0.013 = 26$ kJ. The energy transfer efficiency in firing the shell is therefore:

$$\frac{1.16}{26} \times 100 \cong 4.5\%$$

An efficiency of 4.5% might seem surprisingly low but it should be remembered that the shell is never a good fit in the bore of the mortar tube, there are no gas-tight seals around the shell, and that the shell is not perfectly spherical (or cylindrical).

MINES

Mines, on the other hand, are essentially shells that burst at ground level, the mortar tubes being used to give the projectiles height and direction. They are available in the common shell calibres such as 75 mm and can therefore be fired from a common set of mortar tubes. Indeed, if a shell

Figure 4.3 *Shells and mines light up the night sky above the River Thames in London to celebrate the Spice Girls–Virgin Records contract* (Courtesy of Pains Fireworks Ltd.)

does, for any reason, explode prematurely in its mortar tube the effect is known as 'mining'.

Obviously, with no lifting charge and fewer compartments, the mine is a little simpler in construction than the shell, but the range of effects is equally varied and interesting and includes comets, stars, whistlers, hummers, fountains and serpents.

The modest height achieved with the above effects is especially effective when used in conjunction with salvos of shells or as an alternative to shells in display venues (Figure 4.3) where the height is restricted for any reason.

Chapter 5

Fountains

COMPOSITIONS

Fountains or waterfalls are popular fireworks and range from small devices of only approximately 15 mm in diameter up to professional fireworks of 125 mm or more. They may be fired singly or in groups to provide, as the name suggests, a bright fountain of sparks. If a string of fountains is hung from a rope and fired in the inverted position, as is common practice in the UK, the plumes produce a true waterfall effect which can last for several minutes. Similarly large devices can be angled on lofting poles up to eight feet tall or even mounted on revolving blocks to give an interesting spatial distribution of sparks. With the larger fireworks the blue touch-paper is replaced with a fuse or match that terminates in a depression pressed into the surface of the composition.

Although the compositions used in fountains are usually based on black powder propellant, the sparks that are responsible for the fountain effect originate from other substances within the composition. These substances are known as 'emitters' and it is the physical and chemical properties of the emitters that determine the characteristics of the fountain. Various additives are also used to promote the visual effects or to cheapen the composition.

Broadly speaking, the components of the propellant react to produce hot combustion gases which heat up the particles of the emitter and eject them from the body of the firework. On contact with the air, the hot emitter particles ignite to produce the well known fountain effect as illustrated in Figure 5.1. Typical emitters used in such fireworks include carbon, titanium, aluminium, iron or a magnesium/aluminium alloy. Antimony trisulfide (Sb_2S_3) is also commonly used to enhance the 'glittering' effect in a series of chemical reactions with the gunpowder and aluminium.

But in order to gain a deeper understanding of how fountains work it is necessary to grasp the nettle of atomic theory and also of the electron.

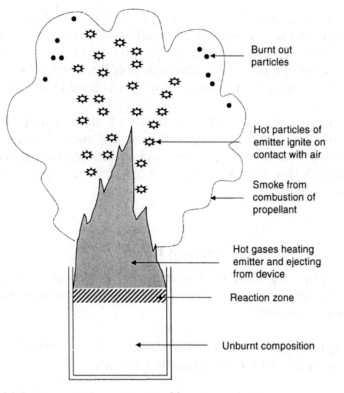

Figure 5.1 *Diagrammatic representation of fountain combustion*

Atomic Theory

Imagine an atom magnified to an immense size (like the Millennium Dome). If you had the magical and perceptive powers of Alice in Wonderland you might penetrate the outer cover of the empty dome and wander through its atmosphere which, upon closer inspection, would be seen to be 'not quite empty'.

Fuzzy little clouds of electrons would make blurred impressions on your imagination and you would feel powerful forces emanating from them. As you got closer to the centre of the dome the forces would continue, until at the centre you would discover the nucleus – a quivering concentration of matter and energy about the size of a grape. On this scale the pips would represent the sub-atomic particles (including protons and neutrons) held together by enormous energy.

Now the protons carry positive charges, while the neutrons are electrically neutral. But nature has arranged things so that the negative charges of the electrons are exactly balanced by the positive charges of the protons, thus keeping the atom 'whole'.

In this magnified perspective the dome is like a vast bubble with the nucleus in its centre. The rest of the atom is sparsely populated but also vibrant and dynamic. The ghostly electrons are arranged in vague clouds and have no clearly defined position. Heisenberg's Uncertainty Principle (1927) tells us that we can't pin-point their positions. Instead, we have to talk in terms of the 'probability' of there being electrons of a certain energy in certain positions (or orbits) around the nucleus at certain times.

The most significant feature from a firework maker's point of view is that the outermost electrons (furthest from the nucleus) posses higher energies than their innermost cousins and are also reactive. In fact, they are so reactive that they can be made to rearrange their positions in the hierarchy of the atom.

Quantum Theory

Profound and astonishing discoveries from the early parts of the 20th century also tell us that the classical laws of physics break down under two extreme conditions:

(1) when things are viewed on the scale of the universe; and
(2) when things are viewed on an atomic scale.

So far as atoms are concerned, the quantum theory (as it is called) dictates that the electrons are arranged in groups (starting with pairs) around the nucleus in discrete energy levels, or shells. As the distance from the nucleus increases, the number of electrons in each shell generally increases, as does the energy.

But there is a limit to the number of electrons in each shell; for example, the first shell can hold up to two electrons, the second has eight, and so on.

Also, the amount of energy ascribed to each shell is fixed in an orderly fashion. The electrons in the first shell may have an amount of energy, x, but no more and no less. Similarly, for succeeding shells, the energy increases by a series of fixed amounts. There is no 'shedding' or gaining of energy in arbitrary or indiscriminate amounts.

When an atom becomes excited (for example, under the influence of heat) an electron might 'jump' to a higher energy level or shell within the atom, but only to a precise energy level or 'step'. Similarly, when an atom 'relaxes', the electron has a natural tendency to occupy the lowest available shell. In doing so it 'steps down' to the lower energy level, shedding a 'quantum' of energy as it does so.

The emission of radiation due to redistribution of electrons among the permitted molecular energy levels follows the same pattern as with atomic emissions.

But for any absorption or emission process, the total energy must be conserved. This leads to the relation given by equation (5.1),

$$E^1 - E^{11} = E = hv \qquad (5.1)$$

where E^1 is the energy of the higher energy state, E^{11} is the energy of the lower energy state, and v is the frequency of radiation that is related to the energy difference, E, by a constant, h, known as Planck's constant.

Thus, a molecule may exist in many states of different energy. The internal energy in a certain state may be considered to be made up of contributions from rotational energy, E_{rot}, vibrational energy, E_{vib} and electronic energy, E_{el} as described by equation (5.2):

$$E = E_{rot} + E_{vib} + E_{el} \qquad (5.2)$$

Electronic, vibrational and rotational changes all contribute to the emission (or absorption) of a single photon but, as will be seen from Figure 5.2, electronic energy levels are widely separated. For a diatomic molecule such as TiO, the electronic energy levels are separated by about $400 \ kJ \ mol^{-1}$, while the vibrational and rotational energy levels are separated by about $20 \ kJ \ mol^{-1}$, respectively.

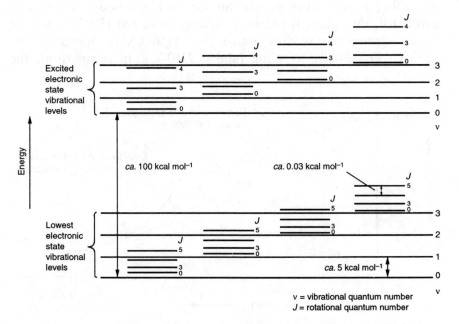

Figure 5.2   *Energy levels of a diatomic molecule (e.g. TiO). Two electronic states are shown, together with some vibrational and rotational levels*

THE COLOUR OF SPARKS

As well as emitting at characteristic frequencies due to specific energy transitions, an emitting species such as a hot metal oxide will also possess a component due to black body radiation, which occurs at all frequencies and is a function of temperature only.

Therefore, the emissivity or radiance of a hot oxide particle will comprise contributions from both black body radiation and molecular energy transitions as illustrated in Figure 5.3.

Although it is known that the colour of black body radiation is only dependent upon temperature, sparks have colours that are also dependent upon the type of emitting material. However, the form of the radiance curves does not relate exactly with known molecular energy transitions. This suggests that the mechanism of emission in excess of black body radiation is not yet fully established. It is possible that some emission bands only become active when the metal oxide particle is molten, or that the energy is dissipated simply *via* collisions with other molecules rather than the emission of photons.

Thus, although the colour of sparks is dependent upon flame temperature and may be similar to that of black body radiation, the overall colour effect can include contributions from atomic line emissions, from metals (seen in the UV and visible regions of the electromagnetic spectrum), from band emissions from excited oxide molecules (seen in the UV, visible and IR regions) and from continuum hot body radiation and other luminescence effects. So far as black body radiation is concerned, the colour is known to change from red (500 °C; glowing cooker element) to reddish orange (510–1150 °C), to orange (1150–2250 °C) and finally to white (above 2250 °C). In comparison, the

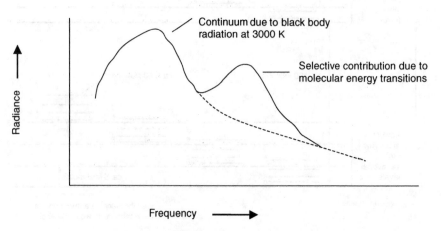

Figure 5.3 *Schematic representation of firework fountain radiance based on titanium emitter*

Table 5.1 *Adiabatic flame temperatures for various elements*

Material	Adiabatic flame temperature (°C)
Be	4000
B	2600
Mg	3200
Al	3500
Si	2300
Ti	2900
V	3100
Mn	3100
Fe	2200
Zr	4200
Mo	2600
W	2700

adiabatic flame temperatures for various elements are as listed in Table 5.1.

Table 5.2 lists some of the common active materials present in fountains. It shows that 'pure' metals such as aluminium or titanium have boiling points that are high enough to enable the particles to survive passage through the flame of the fountain (which is typically around 2700 °C). On the other hand, magnesium has a relatively low boiling point (about 1120 °C) with the result that the metal is volatilised in the flame and good sparks are not produced.

However, on the grounds of cost, availability, reactivity and safety, the list of metal powders used by the firework maker reduces to Al, Ti, Fe and Mg/Al alloy (magnalium).

Table 5.2 *Some active materials present in fountains*

Material	Iron	Aluminium	Titanium	Potassium sulfide
Origin	Metal or chemical reaction	Metal	Metal	Chemical reaction
Composition	Fe (+C)	Al	Ti	$K_2S \sim K_2S_7$
Melting point (°C)	1536	662	1660	143–840 (approx.)
Boiling point (°C)	2872	2493	3318	
Oxide melting point (°C)	1377 (FeO)	2042 (Al_2O_3)	1870 (TiO_2)	
Oxide boiling point (°C)	3417 (FeO)	3527 (Al_2O_3)	3827 (TiO_2)	
Oxygen required for burning (g g^{-1})	0.289 (FeO)	0.890 (Al_2O_3)	0.667 (TiO_2)	
Heat of combustion (kJ g^{-1})	4.9 (FeO)	31.0 (Al_2O_3)	19.7 (TiO_2)	

In practice, the sparks due to an aluminium-based emitter appear yellow-white, whereas titanium gives silver-white sparks and iron filings give a gold effect. For orange-red sparks, potassium sulfide is preferred, the emitter being formed from chemical reactions in a gunpowder type of composition as previously detailed in Table 1.2.

THE BRIGHTNESS OF SPARKS

A further factor that contributes to the overall appearance of a firework fountain is the brightness of the sparks. As with colour, the brightness is dependent upon the temperature and characteristics of the material used. The brightness of black body radiation varies with temperature as shown in Figure 5.4.

This figure shows that the brightness at 1000 °C is assigned the arbitrary value of unity while the scale is normalised to it. Thus the brightness at 2000 °C is 5000 times greater than the brightness at 1000 °C. Conversely, the brightness at 600 °C is only 1/10 000th the brightness at 1000 °C.

PARTICLE COMBUSTION

The primary consideration in the burning of a metal particle in air is the limitation of the temperature attained by the boiling of the resultant oxide.

Two schemes for particle combustion have been proposed which differ mainly in the consideration of the condensed oxide formed by the combustion reaction.

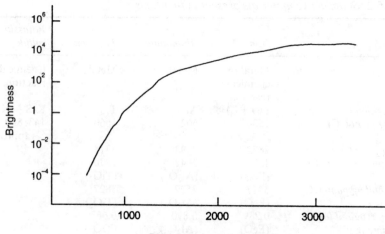

Figure 5.4 *The brightness of black body radiation as a function of temperature*

In the first scheme the metal boiling point is less than the oxide boiling point and the model consists of a vaporising droplet of metal surrounded by a detached reaction zone where condensed oxides appear as fine droplets. The reaction rate is said to be controlled by the vapour phase diffusion of metal and atmospheric oxygen into the reaction zone as in Figure 5.5.

For vapour phase combustion, the burning rate of spherical droplets can be expressed as in equation (5.3):

$$W = Kr^n \tag{5.3}$$

where W is the burning rate, K is a constant involving the latent heat of vaporisation, r is the droplet radius and n is a constant (approximately 1).

In the second scheme, the metal boiling point is greater than that of the oxide and the model suggests that reaction occurs at the metal droplet surface when the vaporised droplet is said to be surrounded by a bubble of molten metal oxide, as in Figure 5.6.

For example, titanium is a non-volatile metal with a melting point of about 1660 °C and boiling point approaching 3320 °C. The oxide TiO_2 has melting and boiling points of 1870 and 3827 °C, respectively. In the fountain, large 'flitters' build up a brittle oxide layer on the surface before any melting occurs. Oxidation continues with heat feedback to the metal until the 'flitter' melts, shattering the oxide shell and fragmenting into small droplets which burn by diffusion at around 3000 °C.

The kinetics that control the small droplet reaction are characterised by the dissolution of titanium oxide which, in turn, exposes further, unoxidised metal. Interestingly, this process appears to be independent

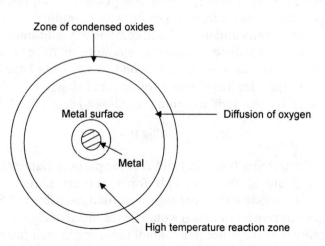

Figure 5.5 *Combustion mechanism for vapour phase diffusion flame*

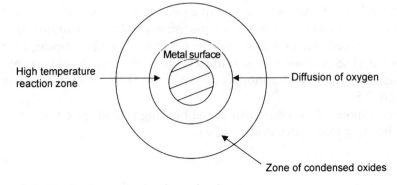

Figure 5.6 *Combustion mechanism for surface burning*

of the type of oxidant, whether it be potassium perchlorate, potassium nitrate or atmospheric oxygen.

Titanium is therefore an important ingredient in fountain compositions. It is characterised as a non-volatile metal with non-volatile oxides. The particles are easily ignited, even in the form of large 'flitters', and once ignited they grow progressively brighter and finally explode in a spectacular star formation.

The fragmentation of sparks has been observed in several metals (including magnesium, aluminium and titanium), and where such fragmentation occurs violently it is termed 'popcorning'. In order to account for this phenomenon, several mechanisms have been suggested.

The first mechanism proposes that metal volatilisation causes rupture of molten droplets (as with magnesium), whereas the second considers the production of a volatile oxide such as CO inside materials such as steels that contain an excess of 0.1% carbon. The third mechanism involves the formation of oxy-nitride compounds which decompose at high temperatures, liberating nitrogen (as with titanium).

The easy ignitability and uninhibited combustion of titanium is largely due to its ability to dissolve sizeable amounts of oxygen as a solid solution without the formation of a second phase. Data have also been published on the size dependence of the burning time of titanium particles in an 'oxygen-rich' medium according to equation (5.4),

$$\log T = 1.59 \log D - 1.30 \qquad (5.4)$$

where T is the burning time (ms) and D is the particle diameter (μm).

Thus for a life of three seconds for a burning particle travelling through air, a particle size in the region of 1000 μm (or 16 BSS mesh) is required. In this respect titanium would be the metal of choice because of its ignitability. On the other hand, aluminium might well prove hard to ignite in such large particle sizes.

The chemical composition as given for the 38 mm fountain shown in Figure 5.7 relies on the gunpowder ingredients potassium nitrate, sulfur and charcoal to provide heat and gas while antimony trisulfide and fine aluminium act as emitters. Barium nitrate is effective in producing intermittent burning and enhances the flickering effect when used in conjunction with Sb_2S_3. The organic substance, dextrin is used as a binding ingredient which helps to consolidate the pressed composition.

Charcoal is used in excess because the decomposition of the extra charcoal is endothermic, the overall effect being to lower the exothermicity of the fountain composition and so reduce the burning rate. In addition, more gaseous products are produced than in the case of the decomposition of KNO_3, *i.e.*, at STP, for every gram of KNO_3 that decomposes, 0.39 litres of gaseous products are produced, whereas for every gram of charcoal that decomposes, 1.3 litres of gaseous products are produced (at STP). However, the main advantage in using extra charcoal is that a reducing atmosphere is produced within the fountain such that the possible reaction of the emitter prior to ejection is greatly reduced.

More exotic effects call for more exotic materials, and considerable effort has gone into formulating compositions that are both spectacular in effect and safe to produce and handle. Thus a 30 mm fountain might contain mealed (or fine) gunpowder, potassium nitrate, sulfur, charcoal, antimony trisulfide, barium nitrate, fine aluminium and flitter aluminium with a dextrin binder. This composition is certainly a good deal more

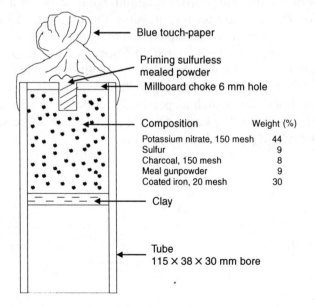

Composition	Weight (%)
Potassium nitrate, 150 mesh	44
Sulfur	9
Charcoal, 150 mesh	8
Meal gunpowder	9
Coated iron, 20 mesh	30

Blue touch-paper

Priming sulfurless mealed powder

Millboard choke 6 mm hole

Clay

Tube
115 × 38 × 30 mm bore

Figure 5.7   *38 mm fountain*

complicated than that used for sparklers but is relatively safe to produce and gives a good burst of white sparks.

The charcoal, or rather the coated charcoal, contributes to the fountain effect as does the gunpowder and aluminium by processes such as those described above. The flitter aluminium has a rather coarser particle structure than does the fine aluminium so that sparks from the former are longer lived and can survive a greater drop-height. Antimony trisulfide is commonly used to enhance the 'glittering' effect in a series of chemical reactions with the gunpowder and aluminium.

Thus, in summary, the composition can be divided into propellant, emitter and additives. The propellant is invariably gunpowder, whilst the emitter might be carbon, steel, iron, aluminium, magnesium/aluminium alloy or even titanium. Additives are often used to promote the visual effects and to cheapen the composition.

Some fountain compositions tend to be oxidant-rich due to the presence of excess potassium nitrate or sometimes various oxalates. The reason for this is to reduce the burning rate and/or to enhance the visual effects. Certainly if gunpowder is considered to be a mixture of fuels (charcoal and sulfur) and oxidant (potassium nitrate) then the maximum rate of burning should coincide with a slightly under-oxidised system. The burning rate is therefore reduced by adding excess nitrate to the system.

Varying the ratios of the components must be exercised with a degree of caution however. With titanium compositions in particular there is a definite limit to the gunpowder/titanium balance beyond which the composition can explode during pressing; the titanium 'flitters' act as tiny razor blades which can friction ignite the gunpowder.

Fountains are particularly suited to festive occasions such as weddings, where silver and gold can be produced alternately from the same fireworks to give graceful arches and fans, or they may be used to supplement other effects such as personalised messages with hearts and flowers. Alternatively, the larger fountains can make successful contributions to daylight pyrotechnics displays.

Chapter 6

Sparklers

There are two main types of firework: wire sparklers that are sold in many shops for most of the year and tubed sparklers that resemble pencils in shape and size.

WIRE SPARKLERS

The wire sparkler might look the simpler of the two types but it can in fact be the more difficult to make. First, a good quality wire must be used that will not corrode during the subsequent dipping operations or on storage.

If iron or steel is used as the spark source it too must be protected from corrosion by coating with a low-viscosity oil such as paraffin. A typical gold sparkler composition contains iron filings, aluminium powder, barium nitrate and dextrin or gum arabic as a binder. The mixture must be of the correct consistency for repeated dipping, and the final drying operation, in currents of warm air, must also be carefully controlled. Just to make things more interesting, the sparklers are not made one at a time, but in huge bundles that are dipped together and then dried. In order to make ignition easier, a priming mix can afterwards be painted onto the tip of the sparkler.

As well as acting as a support for the pyrotechnic composition, the steel wire serves as a heat conductor, promoting the smooth propagation of the pyrotechnic reaction along the sparkler.

A common problem in the production of gold sparklers is the tendency for the $Ba(NO_3)_2$ and Al to react in the wet slurry according to reaction (6.1):

$$16Al + 3Ba(NO_3)_2 + 36H_2O \longrightarrow 3Ba(OH)_2 + 16Al(OH)_3 + 6NH_3 \quad (6.1)$$

This decomposition evolves heat which further accelerates the reaction which is detectable by the smell of ammonia. The reaction rate is increased at high pH and can be effectively controlled by maintenance of a suitable pH using a weak acid such as boric acid (H_3BO_3). Stronger acids would attack the Al powder and Fe filings in the composition.

On ignition, barium nitrate and aluminium react exothermically as in reaction (6.2):

$$10Al + 3Ba(NO_3)_2 \longrightarrow 3BaO + 3N_2 + 5Al_2O_3 \qquad (6.2)$$

The above reaction produces heat and a slight gas pressure which ejects the glowing iron filings to form gold sparks which then cool quickly.

Barium nitrate is superior to either sodium or potassium nitrate with regard to physical stability, while the heat concentration is also higher because barium oxide has better refractory properties than does either potassium or sodium oxide.

A potential disadvantage in large fireworks is the high equivalent weight of barium, but for small fireworks such as sparklers the mass of the composition is not an issue.

TUBED SPARKLERS

For the tubed sparkler the main composition is filled loosely and hence there is no need for a binder. A silver sparkler might be based on potassium nitrate, sulfur, charcoal, steel grit and mealed powder (fine gunpowder) as in Figure 6.1. Again, large batches are made – a typical filling box contains 100 rolled paper tubes packed in square formation for the powder to be sifted in. Obviously, each tube must first be plugged in order to prevent the powder charge from falling straight through, and this is achieved by using wooden plugs which are inserted to about halfway down the tubes. During filling, the bundles of tubes are shaken and more powder is sifted in until each is filled. A priming mix based on grain gunpowder, sulfurless mealed powder and gum arabic is then painted across the mouth of each tube which, when dry, can be finished with a wrapping of blue touch-paper and a label applied.

The ignition process is basically the same for any tubed firework, but when the sparks start to fly, the chemistry becomes very different. Sparks are self-luminous but they need atmospheric oxygen to sustain the high temperature oxidising reaction with the steel particles or other emitters that are present, as described in the previous chapter.

In operation, the mealed gunpowder provides the gas and heat necessary for the combustion of the other fuels and oxidisers that

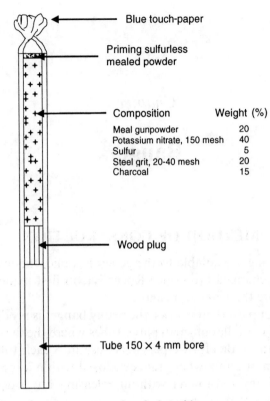

Blue touch-paper

Priming sulfurless
mealed powder

Composition	Weight (%)
Meal gunpowder	20
Potassium nitrate, 150 mesh	40
Sulfur	5
Steel grit, 20-40 mesh	20
Charcoal	15

Wood plug

Tube 150 × 4 mm bore

Figure 6.1 *Schematic representation of a tubed sparkler*

are present. The potassium sulfide that is formed produces orange-red sparks, whereas the steel particles contribute with gold ones. Pine needle-shaped sparks may also be seen when a spark suddenly breaks up into smaller particles. This phenomenon is said to be the result of residual carbon particles exploding in a glowing, active material.

Chapter 7

Bangers

METHOD OF CONSTRUCTION

Although no longer available to the general public, the modest banger (or squib) has changed little since Roger Bacon first made his exciting discoveries more than 700 years ago.

One of the simplest of fireworks, the penny banger is produced in large quantities using bundles of small paper tubes where the first operation is to glue a length of Bickford-type safety fuse into each tube. This fuse contains a train of gunpowder grains enclosed within a rope-like casing which burns from end-to-end without releasing any smoke or flame through the sides of the casing.

The bundle of tubes is then inverted and the explosive charge consisting of fine grain gunpowder and mealed gunpowder is sifted in. Finally, a clay plug is compressed on top of the gunpowder and a mealed powder primer and blue touch-paper are applied to the fuse end.

When the touch-paper is lit, the potassium nitrate, with which it is impregnated, causes the paper to smoulder until it reaches the mealed powder priming. This priming rapidly ignites, which in turn ignites the first grains in the delay fuse which then burn progressively, from grain-to-grain.

The 40 mm or so of delay fuse burns noisily for several seconds and when the end of the fuse is reached, smoke, flame and hot particles are showered onto the main gunpowder filling which promptly explodes. The explosion creates a pressure pulse which operates over a few thousandths of a second but which is enough to shatter the paper tube and enter the air as a blast wave.

It is possible to estimate the pressure that would be reached if the banger as depicted in Figure 7.1 did not explode but remained intact as a 'closed vessel'.

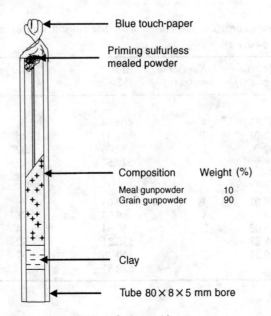

Blue touch-paper

Priming sulfurless
mealed powder

Composition	Weight (%)
Meal gunpowder	10
Grain gunpowder	90

Clay

Tube 80 × 8 × 5 mm bore

Figure 7.1   *Schematic representation of a 'penny' banger*

Volume of Evolved Gases

At standard temperature and pressure (STP) one mole of an ideal gas occupies a volume of 22.4 dm³. Therefore, if the number of moles of gaseous products originating from the gunpowder charge is calculated, an estimate can be made of the 'closed vessel' pressure.

First, a simplified reaction for the gunpowder decomposition can be written as in reaction (7.1):

$$4KNO_3(s) + 7C(s) + S(s)$$
$$\longrightarrow 3CO_2(g) + 3CO(g) + 2N_2(g) + K_2CO_3(s) + K_2S(s) \quad (7.1)$$

From the above reaction the molar quantities associated with the combustion of a 2 g charge of gunpowder can be listed as in Table 7.1.

The gaseous products of reaction are CO_2 (0.012 moles), CO (0.012 moles) and N_2 (0.008 moles). The total number of moles is therefore 0.032 which corresponds to an 'ideal' STP gas volume of 22.4 × 0.032 dm³ or 0.717 dm³.

Theoretical Maximum Gas Pressure

The equation of state (7.2) can now be applied in estimating the maximum pressure in the firework body, assuming the internal volume to be 4 cm³.

Table 7.1 *Molar quantities associated with the combustion of a 2 g charge of gunpowder*

	Molecular mass (M)	Number of moles (n)	Mass (M × n)	Moles per kg	Moles per 2 g charge
Reactants					
KNO_3	101.11	4	404.44	7.68	0.015
Carbon	12.01	7	84.07	13.45	0.027
Sulfur	32.06	1	32.06	1.92	0.004
			520.57 g		
Products					
CO_2	44.01	3	132.03	5.76	0.012
CO	28.01	3	84.03	5.76	0.012
N_2	28.02	2	56.04	3.84	0.008
K_2CO_3	138.21	1	138.21	1.92	0.004
K_2S	110.26	1	110.26	1.92	0.004
			520.57 g		

$$P = \frac{nRT}{V} \tag{7.2}$$

Here, P is the maximum pressure (atm), n is the number of moles of gas, R is the universal gas constant ($0.08205 \ dm^3 \ atm \ deg^{-1} \ mol^{-1}$), T is the estimated flame temperature (K) and V is the volume (dm^3). Therefore,

$$P = \frac{0.032 \times 0.08205 \times 2500}{0.004}$$
$$= 1640 \text{ atm, or } 23\,800 \text{ psi}$$

Of course, the construction of the firework dictates that such outrageously high pressures can never be reached and in the normal course of events the firework body ruptures at a pressure of about 20 atm (300 psi).

Airblast and Sound

On bursting, the banger releases some 5 kJ of energy to the outside world. The noise (or airblast) is perceived as a result of sound waves which originate as an overpressure at the source of the explosion. Sound waves are longitudinal and pass through the air by a rapid interchange between potential energy (pressure) and kinetic energy (motion). Thus the to-and-fro movement of the air molecules is in the same direction as the movement of the waves.

The airblast associated with fireworks can be measured in decibels (dB), since the overpressures involved are comparatively low. The relationship between decibels and bar (or atm) is given by equation (7.3):

$$\text{Pressure in decibels} = 20 \log_{10} \frac{\text{overpressure, in bar}}{\text{reference pressure, in bar}} \qquad (7.3)$$

The reference pressure is usually taken as 20×10^{-11} bar. Thus, for example, if an exploding firework produces an overpressure in the wave front of 0.02 bar (about 0.3 psi), the equivalent in decibels is:

$$20 \log_{10} \left(\frac{0.02}{20 \times 10^{-11}} \right) = \text{or } 160 \text{ dB}$$

As a rough guide, the most vulnerable windows can be broken by 0.1 bar (about 1.5 psi), and so the output by fireworks such as penny bangers is always designed to be less than this.

When we pay two pence instead of one penny for our bangers we expect a bigger bang for our money. One way of achieving this is to include an extra fuel in with the powder mix in order to produce more heat. Commonly, aluminium is used and a typical composition might be based on potassium nitrate, sulfur and pyrotechnic fine aluminium. If the composition is also packed into a larger tube to give, say, double the weight of explosive then the louder bang will be quite noticeable.

In fact, most of the reactions involving the oxidation of aluminium are highly exothermic. The optimum amount of aluminium added to the composition will vary depending upon the amount of available oxygen, but the extra heat generated will more than compensate for the non-gaseous solid aluminium oxide which is formed.

Chapter 8

Roman Candles

METHOD OF CONSTRUCTION

The fireworks that operators refer to as 'candles' or 'Romans' originated in Rome at the time of the Renaissance in the 14th century. Every person in the audience is familiar with these fireworks that eject a succession of stars, mini-shells and hummers into the sky, the projectiles reaching greater heights with every shot while the fireworks send great columns of sparks skywards between those shots.

As with rockets, Roman candles may be fired individually or in fans or bouquets to give a multiple effect (Figure 8.1).

Although candles appear to be among the most simplistic of fireworks they can be quite tricky to produce. The characteristically long paper tubes have very thick walls in order to withstand the high temperatures and internal pressures that operate during firing. A 35 cm long tube, as

Figure 8.1 *Perfectly synchronized bouquets of Roman candles*
(Courtesy of Pains Fireworks Ltd.)

depicted in Figure 8.2, has typically a bore of 15 mm and walls 6 mm thick. It contains up to seven stars that are cylindrical in shape and varied in composition to give, for example, alternate red, green and snowflake effects.

Cylindrical stars are commonly 'pumped' using hand-operated pumps that are forced into the tray of composition. The pump comprises a line of brass or copper tubes, each fitted with a spring-loaded plunger to eject the compressed composition in the form of a star that is of the same diameter as the tube.

A red star might be based on potassium chlorate, strontium nitrate, strontium carbonate and aluminium powder, together with various binders and solvents to facilitate the pumping of solid stars. The composition for a green star could contain barium nitrate, potassium chlorate and aluminium together with binders, while the snowflake star

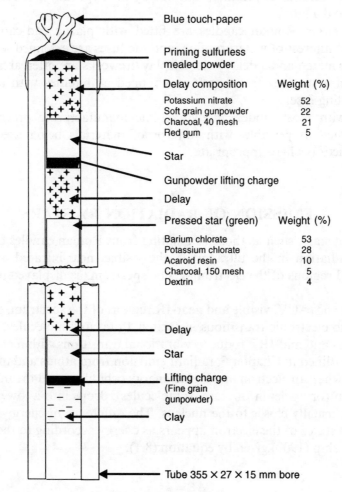

Blue touch-paper

Priming sulfurless mealed powder

Delay composition

Delay composition	Weight (%)
Potassium nitrate	52
Soft grain gunpowder	22
Charcoal, 40 mesh	21
Red gum	5

Star

Gunpowder lifting charge

Delay

Pressed star (green)

Pressed star (green)	Weight (%)
Barium chlorate	53
Potassium chlorate	28
Acaroid resin	10
Charcoal, 150 mesh	5
Dextrin	4

Delay

Star

Lifting charge
(Fine grain gunpowder)

Clay

Tube 355 × 27 × 15 mm bore

Figure 8.2 *Representation of a 27 mm Roman candle*

is commonly made from potassium nitrate, barium nitrate, sulfur, charcoal, aluminium and binders. It is important that the stars are neither too loose nor too tight in the bore of the tube because a loose fit will allow too much gas to escape during ejection and the stars will fail to gain height. Conversely, a very tight fit can result in misfires or explosion within the tube.

In filling, each tube is set on a supporting stud and a quantity of clay is pressed at the bottom using a drift to form a solid plug. A predetermined amount of fine grain gunpowder is then placed into the tube followed by a star. Further gunpowder is then poured into the gap around the side of the star. Delay composition (based on potassium nitrate, sulfur and charcoal) is sifted in and lightly tamped. The process is then repeated, using increasing quantities of lifting charge, followed by a star and delay composition until the tube is full. The filled firework is retained in the upright position for priming, followed by the application of blue touch-paper and a label.

The larger Roman candles are fitted with plastic end-caps in the obvious interest of waterproofing, and the fuses are extended to include a piped match and a delay fuse. Display fireworks are often arranged as fans on a wooden framework or wrapped as bundles with an inter-connecting fuse.

As with most modern fireworks, the manufacturing process is as automated as possible, with waterproof materials being used in the construction where appropriate.

EMISSION OF RADIATION BY STARS

Burning stars, such as those originating from Roman candles or shells, emit radiation in the ultra-violet, the visible, near-infrared and mid-infrared regions of the electromagnetic spectrum, as displayed in Figure 8.3.

In the near-UV, visible and near-IR regions of the spectrum, emission is due to electronic transitions in excited atoms and molecules, while in the near- and mid-IR it is due to vibrational transitions within molecules.

As outlined in Chapter 5, radiant emission from atoms and molecules occurs when an electron in a higher energy orbital around the nucleus of an atom (or nuclei in the case of molecules), drops into a lower energy orbital, usually closer to the nucleus. The difference in energy between the two states of the electron appears as energy according to the Planck relationship (1900) given by equation (8.1),

$$\Delta E = h\nu = \frac{hc}{\lambda} \tag{8.1}$$

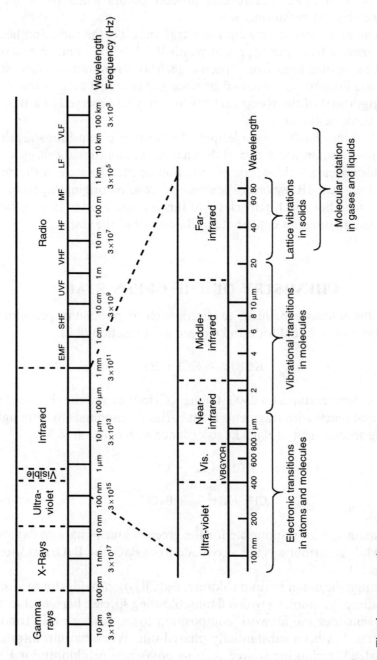

Figure 8.3 *The electromagnetic spectrum*

where ΔE is the difference in energy, h is Planck's constant, v is the frequency of the radiation, λ is the wavelength of the radiation and c is the velocity of light. The reverse process occurs when the atom or molecule absorbs radiation.

In general, excited atoms emit spectral lines, *i.e.* the radiation lies in very narrow wavelength ranges of width 10^{-3} to 10^{-1} nm. In practice, atomic resonance lines from species, such as strontium in a red star, contribute little to the visual effect since the emission falls in the short wavelength part of the spectrum (this line may be observed in a Bunsen burner flame at 461 nm).

On the other hand, for molecules, the electronic transitions result in bands 10–50 nm in width due to the changes in vibrational energy levels which also occur. A third type of radiation emitted by stars in the near-UV–visible–near-IR region is a continuum emission originating from hot particles (*e.g.* hot Al_2O_3 particles) but this is considered to be grey body radiation and does not contribute to the colour of the star.

CHEMISTRY OF THE GREEN STAR

Under the influence of heat, oxidisers such as potassium perchlorate decompose into the chloride and oxygen as in reaction (8.2):

$$KClO_4 \longrightarrow KCl + 2O_2 \tag{8.2}$$

At higher temperatures ($> 2500\,°C$) the KCl ionises and the chlorine that is liberated reacts with fragments from barium compounds to form light-emitting species such as BaCl in accordance with reaction (8.3):

$$KCl \longrightarrow K^+ + Cl^-$$

$$Cl^- + Ba^{2+} \longrightarrow BaCl^+ \tag{8.3}$$

The main species responsible for the green colour of barium flames is BaCl, while contributions are also made from BaO and BaOH as shown in Table 8.1.

Although the use of barium chlorate, $Ba(ClO_3)_2$, provides an oxidising species directly combined with a flame colouring species, barium chlorate greatly sensitises the firework composition towards shock and friction and its use has been substantially phased out. When barium nitrate is used instead, a chlorine source such as potassium perchlorate or PVC [poly(vinyl chloride)] must be added. Hence, a green star composition might contain barium nitrate, potassium perchlorate, aluminium powder and organic binders.

Table 8.1 *Main emission bands/lines for a green star**

Species	Wavelength (nm)	Description
Ba	553.6	Atomic resonance line
Ba$^+$	455.4	Resonance line of first ionised state of Ba
BaOH	**487, 512, 740, 828, 867**	Species responsible for green colour in Ba flames
BaO	535–678	Much weaker than BaOH lines
	549, 564, 604, 649	
BaCl	507–532	Present when composition contains chlorine, and
	514, 524	is main species responsible for green colour

* The wavelengths in bold type indicate the strong emission lines of a green star.

When potassium perchlorate is included in the composition, potassium ions are formed as seen in reactions (8.2) and (8.3). However, potassium emits in the near-IR region of the electromagnetic spectrum and so has little effect on the colour. On the other hand, the ionised form of Ba is undesirable since it emits in the blue region, and potassium salts are often added to Ba stars to suppress ionisation.

Ionisation in Flames

Ionisation of elements in flames is very temperature-dependent, as shown in Table 8.2.

Adding an element of lower ionisation energy can therefore suppress the ionisation of the desired emitting species so that, say, for Ba stars, potassium salts may be added to bring about the change described by reaction (8.4):

$$Ba^+ + K \rightleftharpoons Ba + K^+ \tag{8.4}$$

Table 8.2 *Ionisation of elements in flames*

Element	Ionisation energy (kJ mol^{-1})	% Ionisation at	
		2200 K	2800 K
Li	519.6	<0.01	16.1
Na	494.9	0.3	26.4
K	418.0	2.5	82.1
Rb	402.5	13.5	89.6
Cs	374.9	28.3	96.4
Mg	736.4	0	<0.1
Ca	588.9	<0.1	0.2
Sr	548.4	<0.1	17.2
Ba	502.0	1.0	42.8

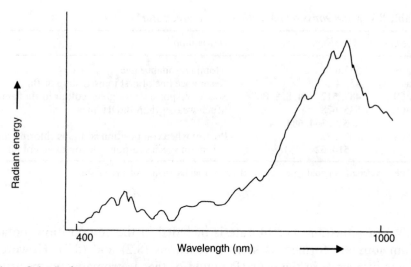

Figure 8.4   *Radiant spectrum of a green star*

Reference to Table 8.1 shows that in the absence of chlorine-containing species the visible emission is dominated by BaOH, in spite of the fact that the equilibrium concentration of BaOH is many orders of magnitude smaller than that of BaO. The reason for this is that the hydroxide is formed directly in an excited state in a process known as chemiluminescence, as shown by reaction (8.5):

$$Ba + OH \longrightarrow [BaOH]^* \longrightarrow BaOH + h\nu \qquad (8.5)$$

Here, [BaOH]* is the excited molecule that releases energy ($E = h\nu$) at frequency ν which corresponds to the green region of the visible spectrum.

Collectively, the molecular and atomic emissions displayed in Table 8.1 give a radiant spectrum as shown in Figure 8.4.

CHEMISTRY OF THE RED STAR

The chemistry that governs barium salts, as used in green stars, is somewhat analogous to that of the strontium salts employed in red stars since both elements are found in the same group of the periodic table. While the use of compounds such as strontium chloride, strontium chlorate or strontium perchlorate might be considered to be appropriate primary ingredients in a red star composition, the salts are ruled out on the grounds of hygroscopicity or mechanical shock sensitivity. Therefore, alternatives such as strontium nitrate are used; the compound serves as both an oxidiser and as a colour source.

Strontium carbonate may be included to enrich the colour, but this substance is not an oxidiser and the 'balance' must be redressed by adding a further oxidiser such as potassium perchlorate together with a balance of fuels such as aluminium powder and organic binders.

During combustion, strontium nitrate and strontium carbonate decompose to give strontium oxide whose spectrum is seen as a pinkish flame due to the positions of the emission bands and to the difficulty in obtaining a high concentration of strontium oxide vapour in the flame. This difficulty is due to the high sublimation temperature of the oxide which is in excess of 2500 °C.

Strontium chloride has a melting point of 870 °C and exerts a considerable vapour pressure above this temperature. The boiling point of $SrCl_2$ is 1250 °C and at temperatures above this it dissociates forming strontium monochloride and chlorine according to reaction (8.6):

$$2SrCl_2 \rightleftharpoons 2SrCl + Cl_2 \qquad (8.6)$$

At still higher temperatures reaction (8.7) predominates:

$$2SrCl + O_2 \rightleftharpoons 2SrO + Cl_2 \qquad (8.7)$$

An excess of chlorine, introduced into reaction (8.7) causes a shift to the left and an improvement in the flame saturation of strontium monochloride. Table 8.3 shows the main emission bands/lines for a red star.

Table 8.3 *Main emission bands/lines for a red star**

Species	*Wavelength* (nm)	*Description*
Sr	460.7	Atomic resonance line
SrOH	**506, 722**	Main species responsible for red colour in Sr flames
	620, 626, 646, 659,	
	668, 682, 707	
SrCl	**618, 636, 661,**	Present when composition contains chlorine
	624, 636, 648, 662,	
	675, 676	
SrO	350–92	Very weak emissions
	750–872	

* Strong bands are indicated in bold type.

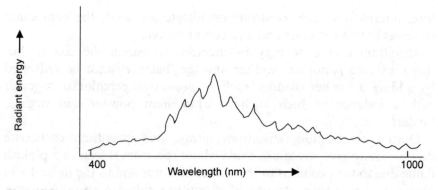

Figure 8.5   *Radiant spectrum of a red star*

Figure 8.5 shows the radiant spectrum of a typical red star.

In practice, both red and green star compositions are formulated to have a negative oxygen balance (*i.e.* there is an oxygen deficiency) since the presence of a reducing atmosphere in the flame inhibits the oxidation of MCl to MO (where M is Sr or Ba), thus enhancing the colour purity of the flame.

Chapter 9

Gerbs and Wheels

GERBS

A gerb (pronounced 'jerb') is a small firework that is built as something of a cross between a fountain and a rocket. Gerbs can therefore be used as 'drivers' for rotating devices or as display pieces with ornamental plumes, or both.

Like the fountain, the gerb, as shown in Figure 9.1, relies on gunpowder to produce thrust, the pressure being increased within the paper case by using a restricting choke or nozzle made of clay. Simple

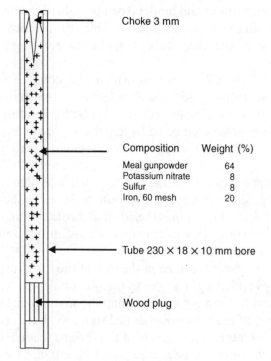

Choke 3 mm

Composition	Weight (%)
Meal gunpowder	64
Potassium nitrate	8
Sulfur	8
Iron, 60 mesh	20

Tube 230 × 18 × 10 mm bore

Wood plug

Figure 9.1 *Schematic representation of a 20 mm gerb*

gerb compositions contain mealed powder, potassium nitrate, sulfur and charcoal – this gives a gold effect.

Method of Construction

In production, gerbs are pressed to varying degrees depending on the thrust required, and have a depression within the choke to produce a rapid pressure rise on ignition. Antimony trisulfide can be used in place of charcoal when a more compact, whitish flame is required.

A vertical rotating device like a windmill might have four sails with driving gerbs at the end of each sail, the central pivot being located at the point where the sails cross. Each gerb is connected *via* a circuit of piped match, which when lit provides more or less simultaneous ignition. For added effect, lances (small flares) can be used to highlight the outline of the sails and these are also simultaneously ignited by inter-connecting piped match. Obviously there is a fair amount of smoke and flame as well as debris generated from the larger devices such as windmills, and the golden rule is to light the fuse (which enters the match) at arms length and then stand well clear.

The effects from gerbs can be enhanced by introducing further ingredients such as steel filings, antimony trisulfide or aluminium to the composition. Small, pressed stars (which are solid pellets containing pyrotechnic composition and binders) can be utilised to add colour to the jet of fire, or a small gunpowder charge might be included on top of the clay in the base of the paper tube to make the gerb finish with a loud report.

'Drivers' are basically gerbs in which the composition has been adjusted to give an increased speed of burning, thereby increasing the thrust. Again, they are used to provide impulse for rotating devices such as wheels and are again connected by lengths of piped match.

The saxon is yet another small firework displaying gerb technology; in this case showing even greater ingenuity in the method of construction and the effects produced. To make this firework, a long tube is filled with gerb-type composition, but plugged at both ends and in the middle using clay. The central, inert portion is then drilled through from the outside to create a pivot hole through which a nail can pass, thus enabling the saxon to revolve like a propeller.

Now if a second hole is drilled in the side of the tube, just beneath one of the end plugs and at right angles to the pivot hole, gas will be released from this second hole on ignition, causing the saxon to rotate.

The other end of the firework is drilled similarly, but making sure that the third hole is on the opposite side of the tube to the second so that the thrust is from the opposite side, as required for circular motion.

In order to achieve simultaneous ignition both holes are connected using match, but the operation time of the firework may be doubled by arranging for one half of the saxon to be ignited first, and then before burning ceases in the first half, a connecting match provides ignition in the second half.

Alternatively, both outlet holes can be bored on the same side of the tube so that as one half burns out, the second portion can be ignited but the saxon will reverse direction, throwing out an astonishing array of sparks.

When used in combination with gerbs, saxons can contribute to spectacular set-pieces such as lattice poles and mosaics, with the gerbs providing overlapping crosses of fire while the saxons fill in the gaps with vivid circles of flame and sparks.

WHEELS

The name Catherine occurs fairly regularly throughout history, most of these ladies being associated with royalty. However, in the third or fourth century there was apparently a virgin martyr of Catholic origin in Alexandria. Legend represents her as being condemned to torture on a toothed wheel; hence the evolution of the firework known as 'St Catherine's Wheel'.

Method of Construction

In the manufacture of Catherine wheels (pin wheels), paper pipe of the length required to make a spiral is closed at one end by twisting or folding it over. Depending on the desired effects, gunpowder with added charcoal and steel is used as a filling to give a glittering shower of sparks of various intensities, while aluminium or titanium can be added to boost the effects. In order to colour the flame, oxidising salts of appropriate metals can be incorporated, but the proportion of gunpowder or mealed powder is usually kept high because, in the absence of any other thrust-producers, the wheel will fail to turn.

After filling, the ends of the tubes are closed and moisture is applied in order to soften the said tubes which are then passed between rollers and partially flattened. Each tube is wound into the well known spiral configuration around a wooden or cardboard disc before placing it in a frame so that the firework can be glued without uncoiling and then being allowed to dry.

A blue touch-paper is applied to the end of the spiral and the central disc drilled to accept a nail or a pin. It is important that anything used to pin the wheel to a post is a good fit in the pivot hole because a tight fit will inhibit rotation, whereas a loose fit will allow the wheel to tip forwards and perhaps catch on the post.

Always a popular firework, Catherine wheels with diameters of up to 50 cm are readily available. Other wheels are more complicated in construction and combine the effects of gerbs, saxons, fountains, lances and rockets in ways that are always attractive to the eye, as well as being dramatic (Figure 9.2).

Giant wheels of fire driven by gerbs are effected by binding the fireworks to the rim of rotating wooden frames. Alternatively, drivers can be sited at the ends of a pair of wooden arms, pivoting from a central block. For added effect, a smaller, contra-rotating pair of arms can be suspended from the same pivot point to provide an inner circle of fire.

Perhaps one of the most ambitious devices is the rocket wheel, which consists of two wheels, each up to one metre in diameter. These are arranged to spin horizontally, rather than vertically, on a specially designed spindle that holds the wheels one above the other, about one metre apart and at a suitable height above the ground.

The rims of the wheels are fitted with eyelets to accept rockets whose sticks are supported by the lower wheel. These are matched to fire at intervals as the wheels revolve, while a battery of Roman candles fires from the centre of the top wheel. Both wheels are fitted with gerbs and drivers to effect rotation, the complete device being appropriately matched such that the overall effect includes two revolving circles of fire, each throwing out glittering arrays of sparks, while candle stars

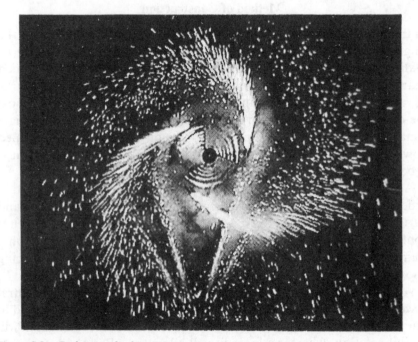

Figure 9.2   *Catherine wheel in action*
(Courtesy of Pains Fireworks Ltd.)

emanate from the centre and rockets fly off in a mad dash in all directions.

By contrast, a crown wheel is one of the simplest fireworks to set up. It resembles a flying saucer and rests, without ceremony, on a nail fixed in the top of a post. Although it spins at about the same speed as a Catherine wheel and uses a similar pyrotechnic composition, the crown wheel is not fixed in any way.

Upon ignition, the firework spins like a Catherine wheel, but in the horizontal plane. This effect is pleasing enough, but the audience is even more impressed when the wheel suddenly takes off like a flying saucer, throwing out a shower of sparks as it goes. The firework has further tricks to play, however, for depending upon the construction, it can dip and rise in the sky once or even twice more before finally burning out.

Chapter 10

Special Effects

Having introduced the basic types of firework, from rockets to gerbs and wheels, it remains to consider the special effects that are essential features of large display fireworks and, indeed, of many of the smaller ones.

Aspects of special effects described in this chapter include fuses (quickmatch and plastic fuse), lances (small coloured flares), set-pieces and devices (assemblies consisting of various types of fireworks linked together), flash, bang and whistle compositions and daylight fireworks (smoke puffs and coloured smokes). Finally, the electrical firing of firework displays is discussed.

QUICKMATCH

Early fuse trains consisting of loose trials of gunpowder were dangerous in that a spark could jump ahead of the advancing flame front and create a second flame front, thus shortening the fuse. The use of quickmatch overcomes this problem somewhat, the gunpowder simply being glued onto a supporting cotton yarn using an adhesive, such as gum Arabic, to form a string-like fuse.

The speed of burning of quickmatch is related to the chemical composition and the mass of the composition on the string. The linear burning rate equation enables a crude estimate of the speed of burning to be made. For example, a typical relationship already seen for gunpowder is given by equation (10.1),

$$R_B = 3.5p^{0.5} \tag{10.1}$$

where R_B is the rate of burning in mm s^{-1} and p is the ambient pressure in psi. Assuming a pressure of one atmosphere or 14.7 psi gives a linear rate of burning of 13.4 mm s^{-1}. This translates to a theoretical burning time of 74 s m^{-1}. In fact, the burning of a typical quickmatch fuse is faster, at around 40 s m^{-1}, this being due to heat effects associated with the construction of the fuse where all three types of heat transfer (radiation, conduction and convection) are present and superimposed on each other.

PIPED MATCH

Whilst quickmatch can be used to form simple fuses for fireworks, both internally and externally, a burning time of some 40 s m^{-1} is too long in instances when a more or less 'instantaneous' ignition is required. Also, quickmatch is rather fragile and tends to kink or lose powder unless it is coated in some way.

Piped match is merely quickmatch that has been enclosed within a paper pipe (Figure 10.1). The paper pipe serves to trap some of the evolved gases and increases the ambient pressure, thereby significantly increasing the rate of burning as well as affording the fuse some protection against mechanical damage.

If one recalculates the rate of burning using equation (10.1) with a revised ambient pressure of, say, 100 psi to reflect conditions inside the paper pipe, a rate of 35 mm s^{-1} is obtained which corresponds to a theoretical burning time of 28 s m^{-1}. But as with quickmatch, secondary effects play an important role, and the actual burning rate rises to at least 4 m s^{-1} (0.25 s m^{-1}), this high velocity being brought about by the chemical composition of the gunpowder, the size of the paper pipe and the tightness of fit of the match in the pipe. For example, it has been shown that a tightly fitting match will result in burn rates that are no quicker than quickmatch in free air. Similarly, slow burn rates are seen if the match is too loose a fit or if the pipe is fired with its ends closed.

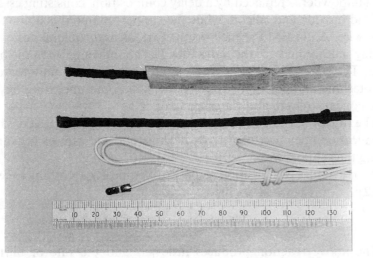

Figure 10.1   *Wirebridge fusehead* (bottom), *quickmatch* (middle) *and piped match* (top)

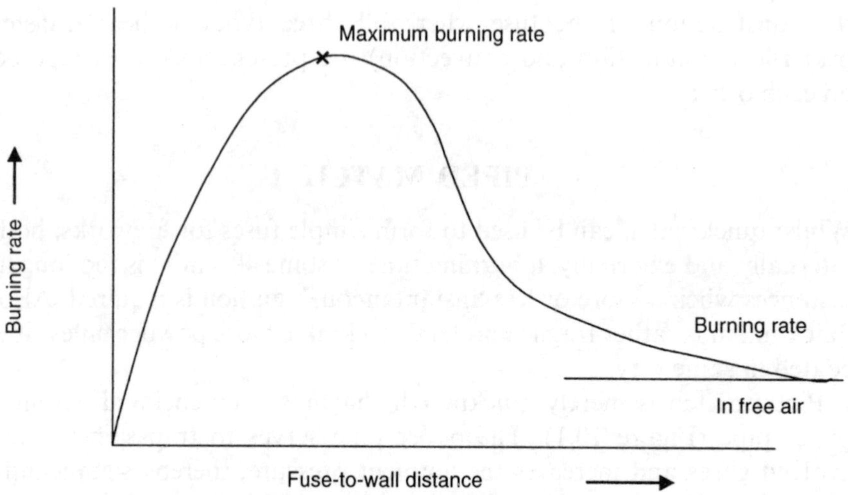

Maximum burning rate

Burning rate

In free air

Fuse-to-wall distance

Figure 10.2   *Diagram showing the variation of the burning rate of quickmatch with fuse-to-wall distance*

Obviously, the flame must be free to travel within the confines of the pipe but at a rate which produces an acceptable pressure increase without bursting the pipe.

Experiments have confirmed that there is an optimum fuse-to-wall distance in piped match where the burning rate is at a maximum, and this effect is presented graphically in Figure 10.2.

PLASTIC FUSE

Plastic Igniter Cord (PIC) is a modern version of quickmatch in which the gunpowder is replaced by a delay composition, consisting essentially of lead dioxide and silicon, while the support material is aluminium or iron wire. A plastic tube encloses the fuse, as in the manner of electrical wiring, but in this case the outer tube has no influence upon the burning rate. The outer plastic serves to colour-code the fuse and contain the pyrotechnic composition, while at the same time imparting flexibility, mechanical strength and waterproofing.

The rate of burning of plastic igniter cord is primarily governed by the chemical composition of the pyrotechnic and the rate of heat transfer along the support wire.

A typical fuse composition reacts exothermically according to reaction (10.2):

$$PbO_2 + Si \longrightarrow Pb + SiO_2 \tag{10.2}$$

The above reaction produces little gas and the rate of burning is relatively free from the effect of gas pressure. However, at high tempera-

Table 10.1 *Grades of plastic igniter cord that are commercially available*

Type	Colour code	Burning speed $(s\,m^{-1})$	Carcase strength
Slow	Green	30	High – iron support wire
Slow	Yellow	33	Nil – no support wire
Slow	Blue	49	Low – aluminium support wire
Fast	Brown	3.3	Nil – no support wire

tures it is quite surprising how much gas can be produced by a composition which is, at first glance, considered to be gasless. The products of reaction in equation (10.2) are both solids at room temperature, but at the reaction temperature of around 2000 °C, significant amounts of vapour are produced.

Plastic fuse is routinely used as a fuse extension to quickmatch or piped match, especially when firing rockets, mines or shells. In the latter case, the piped match fuses from a crate of, say, ten shells connected *via* a single length of plastic fuse which will burn progressively to ignite each of the ten shells in turn.

On the other hand, if a salvo of rockets is required, a fast-burning PIC can be uncoiled into the bottom of a flight-box (which is a box with a wire mesh floor containing rockets, whereby the sticks or tails protrude downwards). On ignition, the plastic fuse rapidly ignites each of the quickmatch fuses protruding from the nozzles of the rockets.

The timing of the above effects is achieved by selecting an appropriate grade of PIC according to types listed in Table 10.1.

LANCES

Lances may be used in conjunction with any of the aforementioned fuses.

A lance is a firework that is about the size of a small pencil that functions in the manner of a flare (Figure 10.3). Thus, the theory of coloured flame production, as presented in Chapter 8, applies equally to the lance.

In a firework lance (for example with a blue flame), we are observing the collective effects of electrons gaining energy and then returning to lower energy levels, shedding energy in the form of photons (coloured light) of certain discrete wavelengths.

Blue is always a difficult colour to produce pyrotechnically because impurities in the chemicals present in the firework tend to produce yellow flames which detract from the blue.

The best blue flames are achieved from compositions based on potassium perchlorate, cuprous chloride, hexachloroethane, polyisobutylene, pyrotechnic copper powder and cellulose dust.

Figure 10.3 *Portfires and lances*

In terms of chemistry, the potassium perchlorate is the oxidiser that oxidises the organic fuels (polyisobutylene, *etc.*) in an exothermic (heat-producing) reaction.

The main species responsible for the blue colour in copper flames is cuprous chloride, CuCl; hence the use of this salt in the composition, together with the chlorine producer (hexachloroethane) and a source of extra copper (pyrotechnic copper powder). The cellulose dust acts as a moderator to control the burning rate of the composition.

Because of the low dissociation energies of copper compounds the flame tends to contain free copper atoms. These emit light in the green region of the spectrum, but on the addition of a halogen (such as chlorine) the colour of the flame changes from green to blue. This change is mainly due to band spectra (of several wavelengths) from gaseous CuCl molecules; (a molecule is a combination of atoms but the processes of excitation are essentially similar).

Initially, the solid CuCl is vaporised. If the resulting gaseous CuCl molecules remain undissociated they will emit the required band spectrum in the blue region.

Increasing the temperature causes dissociation of the CuCl molecules into neutral atoms which, in turn, emit an atomic spectrum composed of atomic (arc) lines. In this state, one of two things can occur. The atoms can combine with hydroxide radicals (charge-carrying ·OH species commonly found in flames) or oxygen atoms to form CuOH or CuO. These gaseous molecules emit a band spectrum and behave like CuCl.

Alternatively, the atoms can be heated further until ionisation occurs (loss of electrons). These ionised atoms emit ultra-violet light and hence a colourless ionic spectrum.

Table 10.2 *Commonly-used colour agents*

Name	Formula	Effect
Barium carbonate	$BaCO_3$	Green, neutraliser
Barium nitrate	$Ba(NO_3)_2$	Green
Barium sulfate	$BaSO_4$	Green
Calcium carbonate	$CaCO_3$	Reddish orange
Calcium sulfate	$CaSO_4$	Reddish orange
Copper(II) carbonate	$CuCO_3$	Blue
Copper(I) chloride	$CuCl$	Blue
Copper(II) oxide	CuO	Blue
Copper(I) oxychloride	$CuCl_2 \cdot 3Cu(OH)_2$	Blue
Cryolite	Na_3AlF_6	Yellow
Sodium disilicate	$Na_2Si_2O_5$	Yellow
Sodium nitrate	$NaNO_3$	Yellow
Sodium oxalate	$Na_2C_2O_4$	Yellow
Sodium sulfate	Na_2SO_4	Yellow
Strontium carbonate	$SrCO_3$	Red
Strontium nitrate	$Sr(NO_3)_2$	Red
Strontium sulfate	$SrSO_4$	Red

It is therefore obvious that the blue lance (in common with other systems) is temperature-dependent. In order to produce a good blue, the temperature must be controlled to ensure that the largest possible amount of vaporised CuCl is present in the flame. A typical spectrum contains wavelength peaks in the region 420–500 nm attributable to CuCl band spectra, a peak at 770 nm due to atomic potassium from the oxidiser, together with CuOH band spectra between 535 and 555 nm.

Other coloured flames follow similar physico-chemical phenomena but operate in different regions of the spectrum. Consequently, the maker of the coloured lance has at his disposal copper salts for blue, strontium salts for red, sodium salts for yellow and barium salts for green, as shown in Table 10.2.

SET-PIECES

The most common set-piece involving lances that one is likely to encounter at a fireworks display consists of a wooden frame measuring about 2 m × 1 m with latticework to support the lances. In fact, the lances are not fixed directly to the lattice but are instead attached to a special form of bamboo that can be softened by soaking it in water and then shaping it to form letters of the alphabet. The bamboo letters are held in position by nailing them to the lattice, after which the letters are picked out using lances that are pinned and glued to the bamboo at intervals. Ideally, the spacing of the lances is about 10 cm. At closer spacing, smoke obscuration and flame overlap can become a problem, whilst more open spacing results in poorly-depicted letters.

Figure 10.4   *Set-piece constructed by the author for the occasion of his daughter's wedding*

The lances are then 'matched' using interconnecting piped match and plastic tape or wire, taking care to ensure that each section of piped match directly above a lance is pricked to create a vent-hole. Otherwise, the force from the ignition of the match can blow it clear of the lance before the firework has been successfully ignited.

Finally, the piped match is fitted with a plastic fuse extension for ease of lighting. In operation, the set-piece is positioned above the ground by lofting on poles 2 m in height. Usually reserved for the end of the display, the set-piece being described here is ignited in a prominent position to display the words 'GOOD NIGHT!'.

Obviously, the number of fire messages and designs achievable in this way is almost infinite – an example is given in Figure 10.4 – and for romantic or other special occasions, outlines depicting hearts and flowers can be set against a lucky couple's initials, while the frame of the set-piece is adorned with gentle glitter fountains of floral bouquets of silver and gold.

For children who are too young to appreciate (or read) love letters in fire, a set-piece called 'the skeleton' might be more entertaining. This firework features the outlines of a skeleton and delights the audience by waving its arms and nodding its head. In this particular case, the lattice work is hinged so that animation can be produced by the pulling of ropes attached to the hinged sections by (unseen) operators.

DEVICES

If the lances are supplemented by other special effects, exotic devices can be made that provide a theme or highlight to a firework display. 'The

Battle' is always a popular device and, as the name suggests, the audience can thrill to the spectacle of ships, tanks or aircraft doing battle across a 40 metre 'no man's land'. The device is, in reality, two devices set the required distance apart with the 'guns' facing each other and the audience watching from one side. Timing of the ignition is critical and two operators must simultaneously ignite the piped match on the two devices. Then it is a case of retiring to a safe distance while the adversaries lob stars and shells and bombs at one another amidst a glorious racket of gunfire and explosions. Usually, the opponent who burns the longest is deemed to be the winner.

Another device that requires a lengthy operating distance is called the 'flying pigeon'. Not too popular with firework operators because of the potential for getting it wrong, a successful firing sees a demented 'pigeon' flying at racing speed across a 40 metre void, shrieking and whistling emphatically until it reaches the other end. The surprise comes when the 'bird' reverses direction and repeats the performance. The errors come if the rope that carries the firework between stout posts becomes too slack or if the firework is lit at the wrong end (the end with the blue dot on it being the 'driver' end).

FLASH AND NOISE EFFECTS

Although gunpowder and its congeners can be successfully used to produce the noise effects of a device such as 'The Battle', there are occasions when a more pronounced flash and bang are required such as, for example, during daylight or when firing a maroon to signal the start (or finish) of a firework display.

In order to produce satisfactory effects a more powerful oxidiser than potassium nitrate is used together with more calorific fuels than sulfur or charcoal. In general, for a given oxidation reaction the heat evolved depends upon the oxidiser anion in the following decreasing order.

$$ClO_4^- > ClO_3^- > NO_3^- > MnO_4^- > SO_4^{2-} > Cr_2O_7^{2-} > CrO_4^{2-}$$

Also, for a given oxidiser anion, copper salts yield more heat than lead compounds, which in turn yield more than sodium, potassium, calcium or barium compounds. In practice, copper salts are not commonly used because of the difficulty involved in their ignition. However, the choice of oxidants and fuels for pyrotechnic effects is extensive, as shown in Tables 10.3 and 10.4.

Most flash and loud report effects are produced by using a relatively powerful oxidiser such as potassium perchlorate (which has a high

Table 10.3 *Properties of inorganic oxidants*

Name	*Formula*	*Molecular weight*	*Density* ($g\,cm^{-3}$)	*Mp* (°C)	$\Delta H_f^{\ominus}$ (kcal mol^{-1})	*Total oxygen* (% wt)	*Available oxygen* (% wt)	*Reaction*
Ammonium perchlorate	NH_4ClO_4	117.50	1.95	d	78.3	54.4	34.2	$2NH_4ClO_4 \rightarrow 3H_2O + N_2 + 2HCl + 2\tfrac{1}{2}O_2$
Ammonium dichromate	$(NH_4)_2Cr_2O_7$	252.10	2.15	d180	420.07	44.4	0	$(NH_4)_2Cr_2O_7 \rightarrow Cr_2O_3 + N_2 + 4H_2O$
Ammonium nitrate	NH_4NO_3	80.05	1.73	170	87.93	60	19.5	$2NH_4NO_3 \rightarrow 4H_2O + 2N_2 + O_2$
Lithium chlorate	$LiClO_3$	90.40		125	87.36	53.1	53.1	$2LiClO_3 \rightarrow 2LiCl + 3O_2$
Lithium perchlorate	$LiClO_4$	106.40	2.43	236	106.13	60.1	60.1	$2LiClO_4 \rightarrow 2LiCl + 4O_2$
Potassium chlorate	$KClO_3$	122.55	2.32	368	89.87	39.2	39.2	$2KClO_3 \rightarrow 2KCl + 3O_2$
Potassium perchlorate	$KClO_4$	138.55	2.52	610	99.24	46.2	46.2	$2KClO_4 \rightarrow 2KCl + 4O_2$
Potassium nitrate	KNO_3	101.10	2.11	334	118.78	47.4	23.7	$4KNO_3 \rightarrow 2K_2O + 4NO + 3O_2$
Sodium chlorate	$NaClO_3$	106.45	2.49	250	82.34	45.1	45.1	$2NaClO_3 \rightarrow 2NaCl + 3O_2$
Sodium perchlorate	$NaClO_4$	122.45		d482	100.60	52.2	52.2	$2NaClO_4 \rightarrow 2NaCl + 4O_2$
Sodium nitrate	$NaNO_3$	85.01	2.26	307	106.6	56.4	28.2	$4NaNO_3 \rightarrow 2Na_2O + 4NO + 3O_2$

d = decomposes.

Table 10.4 *Fuels commonly used in fireworks*

Name	Formula	Notes
Aluminium	Al	
Antimony trisulfide	Sb_2S_3	
Charcoal	C	*ca.* 85% carbon
Graphite	C	
Hexamine	$C_6H_{12}N_4$	Hexamethylene tetramine
Iron	Fe	
Lactose	$C_{12}H_{22}O_{11} \cdot H_2O$	
Lamp black	C	
Magnesium	Mg	
Magnalium	Mg/Al	Typical alloy 50:50
Silicon	Si	
Sodium benzoate	$NaC_7H_5O_2$	
Sodium salicylate	$NaC_7H_5O_3$	
Stearic acid	$C_{18}H_{36}O_2$	
Sulfur	S	
Titanium	Ti	

available oxygen content), together with a finely-divided metal fuel (M) that can react according to reaction (10.3).

$$KClO_4 + 4M \longrightarrow KCl + 4MO \qquad (10.3)$$

Such a reaction is capable of producing over $10 \, \text{kJ g}^{-1}$ in terms of evolved heat.

If one fills the flash composition in a loosely-packed condition into a strong container it will explode violently on ignition, producing a bright flash of light lasting for about a second and accompanied by a loud bang. Pressure build-up accelerates the reaction while rapid energy release produces the flash.

However, a word of caution is needed here because some of the best flashes and bangs are the results of the careful use of some quite powerful and hazardous compositions. For example, the maroon that is often used to signal the start or finish of a fireworks display is loaded with a 'flash' composition containing oxidisers and metallic fuels that must be mixed using remote-controlled machinery.

THE WHISTLE EFFECT

The 'active' ingredients in pyrotechnic whistle compositions are invariably based on aromatic compounds such as gallic acid (10.1) or the salts of aromatic acids including sodium salicylate (10.2) and potassium benzoate (10.3).

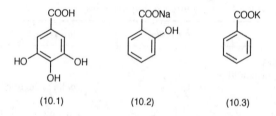

(10.1) (10.2) (10.3)

Originally, salts of picric acid (10.4) were used.

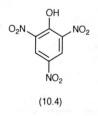

(10.4)

All whistle compositions are hazardous to handle, but those containing picrates are also shock-sensitive with a tendency to explode. Picric acid (2,4,6-trinitrophenol) is, in fact, classed as a high explosive.

In forming whistling fireworks, the aromatic compounds described above are mixed with oxidisers such as potassium nitrate or potassium perchlorate and pressed into tubes. On combustion, a loud whistling sound is produced, whose pitch is related to the length and diameter of the tube.

It is generally agreed that the sound originates from oscillations during burning, when the aromatic compounds create small 'explosions' or 'decrepitation' on the burning surface which results in a change in pressure of the out-streaming gases. A resonating standing wave is created inside the tube whose wavelength increases as the length of tube above the burning surface increases. An increase in wavelength is associated with a lowering of the frequency of the whistle, the effect progressing as the burning composition is consumed and free space increases within the confines of the tube.

SMOKE PUFFS

Bursts or puffs of smoke are produced by explosively-dispersing very fine particles (aerosols), either in the form of combustion products or as inert, solid ingredients. These effects are seen in daylight pyrotechnics displays or when firing 'blanks' from artillery pieces. In the latter instance, a brass case loaded with grain gunpowder is fired from the breech to give white smoke, which is in fact an 'aerosol' of mainly potassium carbonate particles. Alternatively, if a bag of fine powder such as charcoal dust is inserted into the muzzle, together with a small bursting charge of 'flash' powder, a black smoke is discharged.

For coloured smoke puffs, pigments based on metal chromates are utilised, the pigment being intimately mixed with a fuel such as magnesium. On combustion in free air, a smoke cloud is produced that has residual colour due to the chromate.

COLOURED SMOKES

As well as using inorganic pigments in the production of transient smoke effects, a more sustained release can be brought about by using compositions based on gas-producing pyrotechnic ingredients together with readily sublimable organic dyestuffs. In this way, slow-burning 'candles' are produced whose range of colours is limited only by the number of heat-resistant dyestuffs that are commercially available.

One of the simplest, and hence cheapest smoke colouring agents is an azo dye known as 1-(phenylazo)-2-naphthol which is bright orange in colour and also used as a pyrotechnic 'distress' signal. The dye is made by a diazotisation process involving aniline and 1-amino-2-naphthol as shown in Scheme 10.1.

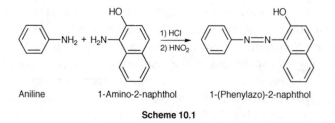

Aniline   1-Amino-2-naphthol   1-(Phenylazo)-2-naphthol

Scheme 10.1

Any gas-producing pyrotechnic compositions must be both hot enough to melt and volatilise the dye and gassy enough to disperse the dye particles. Commonly, a mixture of potassium chlorate (oxidiser) and lactose (fuel) is used in a reaction represented in a simplified form by reaction (10.4):

$$C_{12}H_{22}O_{11} \cdot H_2O + 3KClO_3 + dye_{(s)} \longrightarrow$$
$$dye_{(v)} + 3KCl + 11H_2O + 4CO + 5C + 3CO_2 + H_2 \quad (10.4)$$

For example, the dye, Solvent Yellow 14 has a sublimation temperature of about 125 °C whereas its melting point is 134 °C. The reaction temperature of the chlorate–lactose composition is in excess of 500 °C, which often results in chemical reactions bringing about the destruction of a large proportion of the dye. For example, strong reduction brings about cleavage as shown in Scheme 10.2,

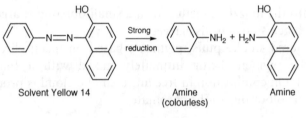

Solvent Yellow 14   Amine   Amine
                                  (colourless)

Scheme 10.2

whereas pyrolysis leads to decomposition as in Scheme 10.3:

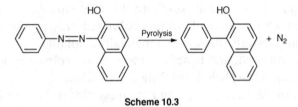

Scheme 10.3

Hence, a typical orange smoke composition contains up to 50% of dyestuff in order to offset losses due to the above reactions.

The loss in colour is due to a loss in conjugation of the aromatic molecule. Dyes absorb light in the visible region of the electromagnetic spectrum by virtue of transitions between electronic energy levels.

A dye appears coloured because one or more components of the light falling upon it are absorbed so that the reflected light is deficient in certain colours. The wavelength of absorption is dependent upon the degree of conjugation (*i.e.* the number of alternate single and double bonds within the molecule).

It can be shown that increasing conjugation causes a shift to longer wavelengths. For example, dyes from orange to green can be made from increasingly conjugated anthraquinones, as presented in Table 10.5.

Although blue dyes usually appear to be blue-black in bulk, a blue colour is seen when they are viewed at the edge of a smoke cloud. This is related to the spectral characteristics, concentration and particle size of the dye.

A cooler-burning pyrotechnic composition based on cellulose nitrate and guanidine nitrate has been developed which produces reaction temperatures of less than 500 °C, thus causing less destruction of the dye.

Cellulose nitrate (10.5) is a common component of propellant compositions because the fuel and oxygen are combined within the same molecule,

Table 10.5 *Variation of colour with increasing conjugation of anthra-quinone dyes*

Material	Colour absorbed	Colour seen
Anthraquinone (colourless)		
1-amino-anthraquinone	Greenish blue	Orange
Disperse Red 9	Blue-green	Red
Disperse Blue 180	Yellowish orange	Blue
Solvent Green 3	Red	Bluish green

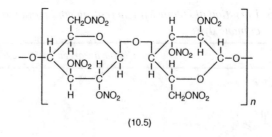

(10.5)

whereas guanidine nitrate (10.6) is rich in nitrogen and is an efficient gas producer.

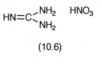

(10.6)

The reaction of the two compounds can be expressed in a simplified form by reaction (10.5):

$$C_{12}H_{14}N_6O_{22} + 7CH_6N_4O_3 + dye_{(s)} \longrightarrow$$
$$dye_{(l)} + 16CO + 25H_2O + 17N_2 + 3H_2 + CO_2 + 2C \quad (10.5)$$

The pyrotechnic composition described above will also contain plasticisers, stabilisers and combustion moderators, thus complicating the overall reaction.

FIRING ELECTRICALLY

Most fireworks can be initiated electrically, and moving to the subject of electric igniters brings us into the 20th century with regard to explosives. The wirebridge fusehead (or electric match as it is more commonly known) consists of a small bead of explosive held on a support which is essentially a pair of conducting foils separated by an insulating wafer as shown in Figure 10.5 (and earlier in Figure 10.1). The foils terminate in a bridgewire (or fusewire) that is embedded in the explosive bead. Relatively few explosives function satisfactorily from the transient hot glow of a bridgewire and the one most commonly employed is the lead salt of an organic substance known as lead mononitroresorcinate (LMNR) (10.7).

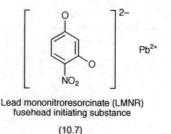

Lead mononitroresorcinate (LMNR)
fusehead initiating substance

(10.7)

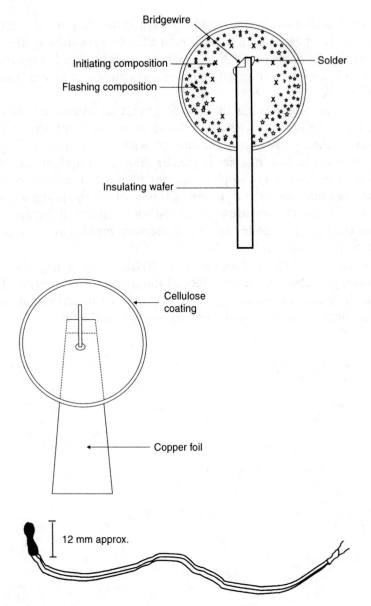

Figure 10.5 *Wirebridge fusehead*

When used in conjunction with more conventional oxidisers and fuels the fast deflagration (burning) from the LMNR ignites the surrounding composition to produce a spurt of flame. Thus the 20 mg or so of explosives in a typical wirebridge fusehead ignite within about 4 ms (depending upon the applied current) and may be used against conventional piped match to ignite shells or any other 'matched' firework.

Provided that the operator has enough connecting wire, a plentiful supply of fuseheads and a firing box, the fuseheads can be connected in series (like the lights on a Christmas tree), each fusehead being securely taped inside a length of piped match to form an ignition train leading to the firework.

Connecting in series has the advantage that the electric circuit can be tested for defective fuseheads or loose connections by doing simple resistance checks. Obviously the current needed to operate a number of fuseheads connected in series is greater than that required for a single device. However, if the applied current is above a certain minimum value, the collective excitation time (during which the bridgewires heat up to the ignition temperature of the LMNR) is shorter than the lag time (when the fusehead ignites and the bridgewire breaks) and a successful multiple firing will result.

The rather complicated setting up of electrically-fired displays is more than compensated for the split-second timing that is achievable. This is especially important when large displays are fired to accompany music, which might be the way of all firework displays in the future.

Chapter 11

Fireworks Safety

The subject of safety is obviously of paramount importance, and this chapter might be used as reference long after the history of fireworks and Vieille's Law have been forgotten.

Fireworks, when not properly handled, can result in injury, loss of life and property damage. The problems of safety are in many ways different from those of most explosives and propellants. In general, the sensitivity of fireworks to shock and impact is considerably less than that of many explosives and propellants. On the other hand, most fireworks are more sensitive to sparks and flame than most other explosive materials.

RADIO HAZARD

When fired electrically, fireworks are connected to wirebridge fuseheads. The fuseheads can inadvertently ignite if subjected to radio hazard (RAD HAZ) when in the vicinity of radio frequency sources such as cellular (mobile) telephones or walkie-talkie radios. Further advice on this should be sought from the makers of the wirebridge fuseheads and of the communication equipment because the characteristics of all these items can vary to some extent.

PUBLIC SAFETY

It can be taken as axiomatic that every person in the fireworks industry recognises that they are dealing with potentially dangerous substances and articles, and that in this regard they will strive to reduce the hazard to themselves and the general public to the absolute minimum.

While it is recognised that there is a continuing emphasis towards large, organised displays, in any free country the inhabitants have the freedom of choice between purchasing and lighting their own fireworks or leaving it to the professionals. It must be said that where fatalities do occur, they tend to be at the hands of private individuals who are

97

handling powerful fireworks recklessly or who are unaware of the dangers.

So far as public safety is concerned, the fatalities due to fireworks in a typical year amount to fewer than 1 in 10 000 000 of the population. On the same terms, this statistic can be contrasted with deaths from infection in hospital (1996) 1000; road deaths (DoT, average 1995–96) 660; accidents in the home (average 1990s) 650; drug-related deaths *ca.* 2000; suicide 86; arson 14; and homicide 130 per 10 000 000 of the population.

Thus, in perspective, the journey by road to any firework display might well be significantly more dangerous than the display itself. However, in order to maintain this high standard of safety it is important to educate, so far as is possible, anybody who is even remotely associated with the handling of fireworks.

First, it must be understood that all fireworks and pyrotechnic compositions are classed as explosives. It is both dangerous and illegal (a) to manufacture gunpowder or pyrotechnic compositions, (b) to assemble such compositions into fireworks, (c) to dismantle (unmake) fireworks or (d) to import fireworks without the appropriate certificate or licence.

Anybody who has responsibility for the handling of fireworks, either privately with a modest sparkler or professionally with gigantic shells, should be aware of the safety aspects as they relate to a particular situation. Generally, if a box or a packet of fireworks is purchased privately there is a list of safety instructions included, or at worst there will be instructions on the paper label of each firework. You should read them because the first line of defence against accidents is information and education. This can be enhanced by adopting a safety philosophy, attending firework operator training courses where necessary, communicating with others, and by planning, example, thought and deed.

In the case of sparklers they should not be handled by small children who might chew at the live composition, poke the wires into their eyes, burn themselves with sparks or pick up the burnt-out wires while still hot. Sparklers should not be handed around while burning, they should not be used near animals, the eyes, flammable materials or other fireworks.

Any remaining sparklers should be kept covered and away from burning fireworks. Anybody who takes all these precautions will have made a safety assessment, either consciously or sub-consciously, and the risk of accident will be minimised.

ORGANISED DISPLAYS

In the broader context we must consider safety as it applies to a long list of topics. For every firework display there is the safety of the general public to consider. Equally important is the safety of the spectators, stewards and attendants, not forgetting the operators themselves. The list should also take account of the safety of animals and property, including buildings, vehicles and personal effects. An ill-aimed rocket can quite easily penetrate a bathroom window and continue burning once inside the room.

A diligent operator must also be concerned with the quality of the fireworks and equipment; that there are no obvious flaws or deficiencies. The frames and timber should be given close scrutiny, as should the conveying vehicle.

Once on-site, the operator must judge the weather. Is it too wet or windy and is the wind in the right direction? Will the sparks from the bonfire blow onto the assembled fireworks? Will the smoke and fall-out be taken towards the spectators?

Communication with the display organiser, hereafter known as the 'customer', should clarify the situation with regard to crowd control, emergency services, first-aid and other amenities. The site must be inspected for obstacles, ground condition, power lines, animals, neighbouring buildings and the fall-out area. If any of these aspects are not acceptable on safety grounds, then the operator has every right to modify or to cancel the display. Obviously, it is helpful if an inspection of the site can be made before the display day, but this is not always possible.

The operator's first point of contact is the customer. After providing proof of identity, essential questions must be asked such as what are the arrangements with stewards, security and emergency services? Are the police and near neighbours aware of the display? Where is the nearest telephone and emergency exit? Are there any first-aid or other facilities?

Inspection of the site and an assessment of the wind direction enables an informed judgement to be made regarding the position of the fireworks in relation to the spectators. If necessary, the positions must be adjusted (or even reversed) so that members of the audience have their backs to the wind so far as is possible. It is absolutely vital that smoke and debris do not drift towards the audience.

Rockets fired vertically will track into the wind and can be carried hundreds of metres down-range before the burnt-out motors and sticks return to earth. Similarly, in high winds, the debris from shells can carry considerable distances and scatter over a wide area. The fall-out area must therefore be inspected, as must the positions of any bonfires on the site.

Assuming that the operator and his colleagues have permission to take their vehicles across the site and that the ground is suitable for vehicular access, an approach can now be made to the firing area.

When the operator looks at the firing area, one of the factors to be considered is the safety distance from the spectators to the first line of fireworks. This is dependent upon the type and size of the fireworks of course, but as an absolute minimum the distance should be between 25 and 50 metres. Ideally, the ground should be soft enough to accept stakes or digging in of the largest mortars, but if not, sandbags must be used. If the ground is not level, the degree of slope must be taken into account when the frames are erected and when the crates of mines and shells are set up. Rough ground can present problems if the display is to be fired electrically when smooth runs of wire are required.

Having inspected the firing area, the fall-out area and the spectators' enclosure, it is time to consult the check-list and to commence the unloading of the fireworks. It is prudent to place 'NO SMOKING' signs within 50 metres of all fireworks, and mobile communication equipment should be banned from the vicinity because of the radio hazard if the display is to be fired electrically.

Trained operators know that the fireworks should be unloaded in small quantities at a time. This not only prevents human ruptures but minimises the danger from inadvertent ignition. Similarly, the 'matching' of shells should be done well away from the main supply of fireworks. Mines and shells are especially dangerous and accidents have been caused by joining pieces of piped match using metallic staples. The correct procedure is to tie the match or fuse in place using string and then to lap the join with tape in order to protect the match against sparks from other fireworks. The shell should be carefully lowered into its mortar tube only after the tube has been positioned ready for firing. Once inside the tube, the head and other parts of the human body should be kept well clear of the muzzle. In the event of a misfire, shells, and indeed all fireworks, should be left for at least 15 minutes before dousing with water. Fuses should never be re-struck if there is the slightest possibility of injury to the operator or others from an uncontrolled ignition.

If a mine or shell is too tight for the mortar tube (assuming that it is of the correct calibre), it should on no account be hammered or crushed to make it fit. Gunpowder is friction-sensitive and it is better to omit the firework from the display rather than have it explode in one's face.

Although professional operators tend to use the same set of mortar tubes on more than one occasion, it is safer to use them once only, particularly during a display because the act of reloading is hazardous if there is any smouldering debris or 'afterglow' from the previous shots within the tube into which a fresh shell is inserted.

In the absence of crates or sandbags, each mortar tube should be partially buried in the ground. In this event, care must be taken that there are no stones or other loose objects in the vicinity that might become missiles in the event of an unscheduled explosion. Having placed all mines and shells at the extreme rear of the firing area it is necessary to ensure that any racks or crates of mortar tubes are positioned perpendicularly or end-wise to the audience so that, in the event of a container tipping over, the contents are not left pointing towards the audience.

Rockets should similarly be located at the rear of the site, along with other aerial fireworks such as crown wheels and maroons, again in positions of safety with respect to the spectators.

Roman candles, set-pieces, wheels and other devices are normally secured to stakes placed in the ground. This is a safe practice so long as the stakes are positioned between the firework and the audience, whereby any loosely-held firework cannot tip towards the audience.

The most favoured knot when tying posts to stakes is the clove hitch. With practice, this knot is quickly effected and easy to undo. However, when tying candles, care should be taken so that the knot is not over-tightened, otherwise crushing of the firework tube might occur which could interfere with the ejection of the stars.

The use of nails in place of rope or string poses additional hazards. Although wooden frames can be securely fixed to posts using nails there is always the problem of protruding nails on dismantling after the display. Perhaps more serious is the danger from flying sparks if a hammer and nails are used, because fireworks are sensitive to sparks as mentioned at the beginning of this chapter.

Bundles of miniature candles packed into circular or square boxes are known as 'cakes' for obvious reasons. They are the simplest of fireworks to set up because one merely places them on level ground, on a waterproof sheet if necessary. However, the fuse that connects all of the tubes runs in a spiral inside the base of the box, and the end that terminates on the outside wall of the box can be difficult to locate (especially in the dark) and be rather slow in operation. Some tens of seconds might elapse before the first tube ignites, but do not be tempted to return to the firework once the fuse is lit. As always. It is a matter of 'light the blue touch-paper' and 'retire immediately'.

Finally, the question of safety can be addressed to portfires. These are the long, pencil-like fireworks that emit a shaft of flame for a few minutes in order to light the fuses of the display fireworks. Obviously, for any display lasting for more than a few minutes, a number of portfires are required. It might be tempting to carry these spares in a pocket but remember the sensitivity hazard. A flying spark could set them off. The correct procedure is to carry one portfire at a time in a gloved hand, along with a box of matches to provide a re-light if necessary. Spares should be stored under cover, at some convenient position in the firing area. The composition inside a portfire is somewhat loosely pressed which means that, on occasion, an air gap can develop in the column of burning ingredients that will be enough to put the flame out. In order to guard against this the portfire is tapped (gently) in a 'nose down' position to close any gaps before lighting.

After the display, any burning embers should be doused with water, while fireworks that have failed in any way should be segregated from spent fireworks, labelled as MISFIRES and packaged separately. On no account should spent cases or live or faulty fireworks be thrown onto the bonfire. Mortar tubes should be inspected for glowing embers, which should be extinguished before transporting.

Any remaining portfires or other live fireworks should be returned to their original packaging, and ideally packaged so that they cannot rattle or shift about in transit. Any spilled powder should be washed away with copious quantities of water.

Enjoy your fireworks but remember that failure to act safely and responsibly can lead to accidents; just as it will with the ubiquitous garden barbeque. The motor vehicle is still ahead on points though – in 1997 there were over 65 000 fires in vehicles resulting in some 100 deaths.

Chapter 12

Fireworks Legislation

THE EXPLOSIVES ACTS

The important subject of legislation applied to the explosives industry, including fireworks, commenced in the form of the Explosives Acts of 1875 and 1923.

Nowadays, regulatory provisions are also applied, and those governing health and safety fall into four categories: (a) Acts of Parliament, (b) Regulations made under those Acts, (c) Codes of Practice, and (d) Guidance Notes.

When a Bill has been debated in Parliament and received Royal Assent it becomes law. Regulations made under the Acts also become law, and so the Factories Act of 1961 and the Health and Safety at Work Act of 1974 must be obeyed because they are part of criminal law.

The fireworks industry in the UK is too small to warrant continuous attention by Parliament, but two corporate bodies have been established in accordance with the Health and Safety at Work Act that control safe practices within factories, including fireworks factories, and this, in turn, ensures a safe standard of products that reach the general public.

THE HEALTH AND SAFETY COMMISSION

The Health and Safety Commission (HSC) consists of a Chairman appointed by the Secretary of State for Employment, and not less than six and not more than nine other members drawn from industrial organisations, local authorities and professional bodies.

THE HEALTH AND SAFETY EXECUTIVE

The Health and Safety Executive (HSE) is a statutory body co-ordinated by a Director General and two other people appointed by the Commission.

Health and Safety Inspectors (including Explosives Inspectors) are appointed by the HSE to ensure compliance with the relevant statutory provisions. The inspectors have very wide ranging powers, and this is another factor that contributes to fireworks safety.

A welcome addition to the controls imposed by the HSE which relate to the manufacture, keeping, safe conveyance and importation of explosives, including fireworks, is the British Standard BS7114. This standard applies specifically to fireworks and was promulgated in 1988 in order to deal with aspects such as the categorisation, performance testing, quality control and labelling of all types of fireworks.

Amazingly, no such protocol existed in the UK before 1988, when it was difficult to prevent dangerous, imported fireworks from reaching the shop shelves. Thus, unsuspecting consumers could, in theory at least, purchase items containing prohibited mixtures of, say, sulfur and potassium chlorate and drop them onto the back seat of the family saloon where they could ignite without warning.

Anybody who studies the legislation will see that BS7114 makes a notable addition to the Explosives Acts of 1875 and 1923, the Fireworks Act of 1951, the Health and Safety at Work Act of 1974, the Consumer Protection Acts of 1978 and 1987, and the Classification and Labelling of Explosives Regulations of 1983.

Hitherto, imported fireworks were subject to licencing and a 'spot check' by the HM Explosives Inspectors, while labelling was evaluated on the basis of common sense and past experience.

On the other hand, fireworks manufactured within the UK were controlled by factory licencing and statutory inspections by the Explosives Inspectors.

In the absence of any approved official standard, procedures for quality control, overseas manufacture and importation were tenuous to the extent that it was often unclear as to whether some fireworks were suitable for sale to the general public or whether they should be used by a licensed importer or manufacturer.

A published specification was thus called for in the form of a British Standard.

BRITISH STANDARD FOR FIREWORKS

British Standards are created by a committee that includes representatives from those involved in the manufacturing industry, and in this case the result was BS7114 which is divided into three parts: Part 1, Classification of Fireworks; Part 2, Specification of Fireworks; and Part 3, Methods of Test of Fireworks.

In Part 1, fireworks are divided into four categories that relate to where the products are used and the level of associated hazards.

—Category 1 fireworks are suitable for use inside houses and include toy caps, party 'poppers', cracker snaps, throwdowns, smoke devices and sparkers (both hand-held and non-hand-held). They are often used by children under adult supervision and are either hand-held or designed to function quite close to a person. Factors such as the flammability of clothing materials and other household effects and the possible injury, particularly to the eyes, face and hands, are taken into account when assigning a firework to Category 1. The potential hazards to consider include excessive flame or explosion, hot particles, hot slag and sharp fragments.

—Category 2 fireworks are intended for outdoor use in confined areas, such as small gardens, where spectators are expected to be at a distance of at least five metres when the firework is functioning. The person lighting the firework is also expected to retire to a safe distance and so the fusing of the firework (blue touch-paper) becomes important. Examples of fireworks in this category include fountains, Roman candles, mines, wheels, rockets and large sparkers (both hand-held and non-hand-held). The main hazards associated with Category 2 fireworks are ejected debris, burning matter and erratically-flying rockets.

—Category 3 fireworks are designated for outdoor use in large open spaces such as parks, sports fields and open land, and spectators are expected to be at least 25 metres away when the firework is functioning. This category includes the proportionately larger fountains, Roman candles, mines, wheels, rockets, non-hand-held sparklers and combinations such as devices. Again, the main hazards are ejected or burning matter and erratically-flying projectiles.

—Category 4 fireworks are of such size that they are not intended for use by the general public and therefore do not feature significantly in BS7114.

Part 2 of the British Standard lists all of the main types of firework within Category 2 and Category 3, and provides specification requirements by which the construction and performance of these fireworks can be controlled. Aspects that are addressed include means of ignition, projection of burning matter, projection of debris, principal pyrotechnic effects and the angle of flight of rockets.

The important subject of labelling is addressed in some detail because Categories 1, 2 and 3 all have labelling requirements that apply to the packaging as well as to individual fireworks. In general, the phrases on the label are selected from standard lists that ensure conformity and unambiguity. Thus, the labelling must be in English, in letters of a specified type and size, and all fireworks must be labelled with their intended use and type name. The auxiliary verb 'must' is chosen deliberately here to express necessity or obligation. A correctly labelled fountain in Category 2 will have 'GARDEN FIREWORKS', 'FOUNTAIN', together with the appropriate warnings, instructions and effects, followed by the name and address of the manufacturer or importer. Any fireworks that are deficient as regards labelling should not be purchased. The words 'complies with BS7114: Part 2, 1988' may also be present on the label.

Part 3 of BS7114 is concerned with the methods of testing of fireworks and certain items of auxiliary equipment such as rocket launchers. In essence, the standard calls up chemical and physical test methods that are applied to all types of firework, from toy caps to large rockets. Fireworks that are incomplete or not intended for sale to the general public are excluded.

Physical testing includes a primary examination of the packaging and the fireworks within. A note is made of the general condition of the firework and the presence of any defects such as loose powder. The firework is weighed and, if required, the weight of explosives inside is also determined. In performance testing, the firework is made to function and a note is made of effects such as the fuse burning time, the number of explosions, the ejection of incandescent matter or debris and the attitude of the firework whilst functioning.

Chemical testing is carried out in an approved laboratory because the firework must first be dismantled. 'Wet' methods of analysis are applied that involve analytical grade reagents to detect, in particular, the presence of chlorates in admixture with elemental sulfur. Sulfur–chlorate mixtures are banned in the UK, and a major use of sulfurless gunpowder is in fireworks where chlorates are also present.

UK LIST OF CLASSIFIED AND AUTHORISED EXPLOSIVES

Fireworks that meet the requirements of BS7114 must also be included in a publication called the UK List of Classified and Authorised Explosives (LOCAE). This list originates from the Explosives Acts and relates to civilian rather than to military explosives (details of the latter are UK-restricted). Thus, the 1994 edition of LOCAE contains articles ascribed to Astra, Kimbolton, Pains, Standards, Wells, Hunts, Le Maitre and Nationwide. For example, included in the FIREWORKS columns, will be found STANDARD – SET PIECE. For this particular entry, the United Nations Serial Number is given as 0336, the Hazard Code is 1.4G, the UK Class and Division is 7.2, while the Competent Authority (HSE) Reference is GB 72411.

What do all of these terms mean? Well, the UN serial number 0336 applies to a particular group of fireworks, and acts as an aid to international identification. The HSE or MoD, as part of the process of classification of a product, selects the most appropriate of these four-digit numbers, taking cognisance of the hazard classification code and security requirements.

Upon classification, all explosives (including fireworks) are assigned to dangerous goods Class 1 in accordance with the Classification and Labelling of Explosives Regulations 1983 (CLER). Class 1 explosive substances and articles are then further classified according to their respective hazards. There are five sub-divisions (Division 1.1 to Division 1.5) running from 'mass explosion' to 'no significant hazard'. Fireworks are mainly classified as Hazard Division 1.4 (meaning substances and articles that present no significant hazard, *e.g.* small displays) or Hazard Division 1.3 (meaning substances and articles which have fire hazard and either a minor blast or minor projection hazard but not a mass explosion hazard, *e.g.* large displays). This explains the 1.4 Hazard Division of the STANDARD – SET PIECE, but what does the 'G' mean?

Well, the UN also acknowledge that the safety of substances and articles is best assured by keeping each generic type separate. This is not always practicable, however, and if there has to be a degree of mixing of various explosive substances and articles, then the extent of such mixing is determined by the compatibility of the explosives. Explosives are considered to be compatible if they can be transported (or stored) together without significantly increasing either the probability of an accident or, for a given quantity of explosives, the magnitude of the explosion.

The UN Compatibility Groups are represented by letters from A to S, but omitting I, M and O.

Compatibility Group G comprises any substance which is an explosive substance because it is designed to produce an effect by heat, light, sound, gas or smoke, or a combination of these as a result of non-detonative, self-sustaining, exothermic chemical reactions, or an article containing such a substance or an article containing both a substance which is explosive because it is capable by chemical reaction in itself of producing gas at such a temperature and pressure, and at such a speed, as could cause damage to surroundings and an illuminating, incendiary, lachrymatory or smoke produc-ing substance (other than a water-activated article or one containing white phosphorus, phosphide or a flammable liquid or gel).

That is what 'G' means, and I hope you will agree that it applies to fireworks!

The UK Class and Division for fireworks is 7.2. This should not be confused with the UN dangerous goods Class 1 to which all explosives are assigned.

Class 7, Division 2 relates to manufactured fireworks consisting of any classified explosive and any classified firework composition, when such explosive or composition is enclosed in any case or contrivance, or is otherwise manufactured to form a squib, cracker, serpent, rocket (other than a military rocket), maroon (including signalling maroons), lance, wheel, Chinese fire, Roman candle, or other article specially adapted for the production of pyrotechnic effects, or pyrotechnic signals or sound signals. Provided that a substantially constructed and hermetically closed metal case containing not more than 1 lb of coloured, fire composition of such a nature as not to be liable to spontaneous ignition shall be deemed to be a manufactured firework.

Class 7, Division 1 relates specifically to firework compositions (such as flash powder) and these are referred to as 'substances' rather than manufactured 'articles'.

Other regulations control the packaging and storage of dangerous goods, the transportation by sea, air, rail and road, the training of drivers, the licensing of explosives factories and magazines, and limitations on quantities. Thus the general public are free to purchase certain types of fireworks or do-it-yourself display kits, without licence, from reputable suppliers.

In summary, fireworks legislation is necessary and is designed to prevent accidents and to reduce risk to the absolute minimum. To this end, the HSE helps to ensure:

(1) sufficient knowledge of the properties of compositions being manufactured and hence of the precautions which should be taken;
(2) adequate working instructions and safe systems of work;
(3) planned maintenance of equipment, facilities and buildings;
(4) compliance with advice given by HSE inspectors.

While compliance with a British Standard does not of itself confer immunity from legal obligations, the net effect to the consumer or spectator has been a reassuring increase in the standard of safety of both the fireworks and the people responsible for them. The current trend towards organised displays has not been at the expense of the availability of garden or indoor fireworks, however, and the increasing legislation is mainly the concern of those in the industry who are involved with the importation, manufacture, retail or safe operation of fireworks.

RECENT LEGISLATION

More recently, the Fireworks (Safety) Regulations of 1997, on Consumer Protection, has prohibited the supply of certain fireworks to the general public. These include bangers, fireworks containing bangers, aerial shells and maroons, and shells- or maroons-in-mortars. However, such fireworks are still available to professional operators for firing at organised displays.

The Carriage of Dangerous Goods (Amendment) Regulations of 1999 provide an update on the Classification, Labelling, Packaging and Carriage of Dangerous Goods Regulations (including explosives) published between 1983 and 1997.

Further details of the current legislation are included in the Bibliography that follows this chapter, while full details are available from HMSO Books.

Bibliography

T.L. Davis, *Chemistry of Powder and Explosives*, John Wiley and Sons Inc, New York, 1943.

G.W. Weingart, *Pyrotechnics*, Chemical Publishing Co Inc, New York, 1947.

H. Ellern, *Military and Civilian Pyrotechnics*, Chemical Publishing Co Inc, New York, 1967.

J.H. McLain, *Pyrotechnics*, The Franklin Institute Press, Philadelphia, 1980.

R. Meyer, *Explosives*, Verlag Chemie, Weinheim, 1981.

T. Shimizu, *Fireworks from a Physical Standpoint*, Pyrotechnica Publications, Austin, TX, 1981.

Fireworks, British Standard Parts 1, 2 and 3, BS 7114, British Standards Institution, Milton Keynes, 1988.

J. Quinchoun and J. Tranchant, *Nitrocelluloses*, Ellis Horwood Limited, Chichester, 1989.

K.A. Brauer, *Handbook of Pyrotechnics*, Chemical Publishing Co Inc, New York, 1974.

D.R. Stull and H. Prophet, JANAF Thermochemical Tables, Second Edition, US Bureau of Standards, US Dept. of Commerce, Washington, 1971.

G. Hussain and G.J. Rees, in, 'Proceedings of the 14th International Pyrotechnics Seminar', Jersey, Channel Islands, 1989.

R.J. Rapley and B.J. Thomson, in, 'Proceedings of the 14th International Pyrotechnics Seminar', Jersey, Channel Islands, 1989.

Engineering Design Handbook, Military Pyrotechnics Series, Part 1 – Theory and Application, US Army Material Command, AMCP 706-185, 1967.

J.C. Cackett, T.F. Watkins and R.G. Hall, *Chemical Warfare, Pyrotechnics and the Fireworks Industry*, Pergamon Press Ltd, London, 1968.

J.D. Blackwood, 'The Initiation, Burning and Thermal Decomposition of Gunpowder', *Proc. Roy. Soc. London*, **213**, 1952.

M.E. Brown and R.A. Rugunanan, 'A Thermo-analytical Study of the Pyrotechnic Reactions of Black Powder and its Constituents', *Thermochim. Acta*, 1988, **134**, 413.

R.A. Whiting, *Explosives and Pyrotechnics*, Franklin Institute Research Laboratories, Philadelphia, 1971, vol. 4, nos. 1, 2 and 3.

R.A. Sasse, *The Influence of Physical Properties on Black Powder Combustion*, US NTIS Report AD-A100273/2, 1981.

M.A. Williams, 'Modern Rack and Mortar Designs for Professional Fireworks Displays', *J. Pyrotechnics*, Issue No. 2, 1995.

Rev. Ronald Lancaster, MBE, *Fireworks Principles and Practice*, Chemical Publishing Co Inc, New York, 3rd edn., 1998.

Subject Index